GREEN OSLO

The editors would like to thank the following institutions for their financial support in the preparation for the writing of this book:

The Norwegian Research Council
Oslo City Council
Centre for Development and Environment, University of Oslo
Department of Communication, Culture and Languages, BI Norwegian Business School
Department of Landscape Architecture and Spatial Planning, Norwegian University of Life Sciences
Department of Literature, Area Studies and European Languages, University of Oslo
Department of Sociology and Human Geography, University of Oslo

The editors would also like to thank their fellow committee members for their contributions in time, effort, and advice that led to the writing of this book:

Anne-Karin Thorén
Karen V. Lykke Syse
Mark Brown
Guttorm Grundt

Green Oslo
Visions, Planning and Discourse

Edited by
MARK LUCCARELLI and PER GUNNAR RØE
University of Oslo, Norway

Routledge
Taylor & Francis Group
LONDON AND NEW YORK

First published 2012 by Ashgate Publishing

2 Park Square, Milton Park, Abingdon, Oxon OX14 4RN
711 Third Avenue, New York, NY 10017, USA

Routledge is an imprint of the Taylor & Francis Group, an informa business

First issued in paperback 2016

British Library Cataloguing in Publication Data
Green Oslo : visions, planning and discourse. -- (Urban planning and environment)
1. Environmental policy--Norway--Oslo. 2. Sustainable urban development--Norway--Oslo. 3. Open spaces--Norway--Oslo.
I. Series II. Luccarelli, Mark. III. Røe, Per Gunnar.
307.7'6'094821-dc23

Library of Congress Cataloging-in-Publication Data
Green Oslo : visions, planning and discourse / [edited] by Mark Luccarelli and Per Gunnar Røe.
p. cm. -- (Urban planning and environment)
Includes bibliographical references and index.
ISBN 978-1-4094-3896-0 (hardback)
1. Urbanization--Environmental aspects--Oslo. 2. Sustainable development--Norway--Oslo. I. Luccarelli, Mark. II. Røe, Per Gunnar.
HT384.N62O85 2012
307.7609482'1--dc23

2012005202

ISBN 978-1-4094-3896-0 (hbk)
ISBN 978-1-138-24677-5 (pbk)

Contents

PART III PLANS, POLITICS AND OUTCOMES

List of Figures and Tables

Tables

Tables

List of Contributors

Timothy Beatley is the Teresa Heinz Professor of Sustainable Communities in the Department of Urban and Environmental Planning, School of Architecture at the University of Virginia, where he has taught for the last 24 years. He has written more than 15 books on the subjects of sustainable cities and place making, including most recently *Biophilic Cities: Integrating Nature into Urban Design* (Island Press, 2010). Beatley writes a regular column for *Planning Magazine*, called *Ever Green*, about environmental and sustainability matters. He also collaborated recently on a documentary film about green cities, entitled *The Nature of Cities*, which was broadcast on PBS stations around the US.

Mark Brown is Associate Professor at the Norwegian School of Management (BI), Oslo. He has Masters degrees in engineering from Imperial College, London and in English from the University of Oslo, and has worked in both business (UK) and academia (Norway). He successfully defended his PhD: *Managing Nature – Business as Usual: Patterns of Wording and Patterns of Meaning in Corporate Environmental Discourse*, in June 2008. It is available online at: http://home.bi.no/fgl00083/.

Fredrik Chr. Brøgger is Professor Emeritus of American Literature and Civilization at the University of Tromsø, Norway. He wrote his doctoral dissertation on the interrelationship between literary modernism and US culture in the 1920s and has published widely on American literature and culture, including a book on the theory of (American) culture studies. In recent years, nature writing, ecocriticism and arctic studies have served as his main fields of interest, spawning articles on works by American romantic writers as well as by Ernest Hemingway, William Faulkner, John Dos Passos, Knut Hamsun, Robinson Jeffers, Gary Snyder, Annie Dillard, and Knud Rasmussen.

Eva Falleth is currently Head of Department for Landscape Architecture and Spatial Planning at the Norwegian University of Life Sciences. She joined the staff as a Professor in 1997. She was Fellow and later Professor at the Norwegian University in Science and Technology (1987–1993) and became a research fellow at the Norwegian Institute for Urban and Regional Research in 1994. Falleth has been head of the national board for planning education since 2009, and is a national representative of the Association of European Schools of Planning (AESOP). She is also a member of the editorial board for the Norwegian journal *PLAN*, the editorial board for *Progress in Planning* and member in a research board in the

Norwegian Research Council. Her research focuses on public policy-making in urban and environmental planning. She also has a long-standing commitment to research training and is currently the primary supervisor for three PhD students.

Clas Florgård is Professor Emeritus at the Division of Landscape Architecture, Department of Urban and Rural Development, Swedish University of Agricultural Sciences in Uppsala. In the 1970s and 1980s, he was a consultant in landscape planning and design. During the 1980s, he ran a company of his own, and became Professor of Landscape Architecture in 1994. His research focus has mainly been planning and design for the preservation of existing natural vegetation as part of the future green infrastructure when new urban areas are developed.

Guttorm Grundt held the position of Environmental Affairs Coordinator of the City of Oslo from 1997 until he retired in 2010. He graduated in Architecture from Oxford Brookes University in 1970. He completed the Norwegian Ministry of Environment's upgrading course in planning in 1973, and a senior course in planning at Nordplan, University of Stockholm in 1983. Grundt has comprehensive professional experience, including cooperation with Atelier Parisien d'Urbanisme, Norwegian Institute of Urban and Regional Research, Norwegian Association of Local and Regional Authorities and Scandia Consult, AS. He has been an adviser to the Ministry of Environment on a program for environment-friendly local development, and published articles in professional journals.

Jon Guttu trained as an architect at the Norwegian Institute of Technology (1968), and obtained a PhD from the Oslo School of Architecture (2003). He has worked primarily in the area of physical housing research, first at the Norwegian Building Research Institute, and from 1992 at the Norwegian Institute for Urban and Regional Research (NIBR). His work has included floor plan studies of flats to qualities in the built environment and urban setting. His starting point has been the needs and behaviour of users, and his research method has often comprised interviews and technical analysis. In his doctoral dissertation, 'The Good Dwelling', Guttu described and analysed how technical discussions among housing planners changed from the 1950s to the end of the millennium. The intimate knowledge of the history of Norwegian housing resulting from this work has made Jon Guttu a highly sought-after lecturer.

Peter Hemmersam is Associate Professor at the Oslo School of Architecture and Design. He is trained as an architect and is a former partner in the architectural firm Transform. His main research interest lies in the field of urban design and ecological urbanism.

Per Hedfors is Assistant Professor of Theory at the Division of Landscape Architecture, Department of Urban and Rural Development, Swedish University of Agricultural Sciences in Uppsala. His research focuses on finding the means to challenge the hegemony of vision in landscape architecture and related fields. In his thesis *Site Soundscapes: Landscape Architecture in the Light of Sound*, he developed Sonotope Design Strategies. He has also done research in the development of landscape architecture in the Nordic countries, including analysis of the discipline's development through theories of maintenance, planning and design. As a practitioner, he has been the director of comprehensive planning at a local authority, and has won a competition in playground design.

Karsten Jørgensen is Professor of Landscape Architecture at the Department of Landscape Architecture, Norwegian University of Life Sciences (UMB), and is a founding editor of *JoLA—Journal of Landscape Architecture.*

Mark Luccarelli is Associate Professor in the Department of Area Studies, Literature and European Languages at the University of Oslo. He is the principal organizer of the Nordic Network for Interdisciplinary Environmental Studies (NIES), and the author of *Lewis Mumford and the Ecological Region* (1995) and several articles on space, landscape and representations of nature.

Inger-Lise Saglie is Professor at the Department of Landscape Architecture and Spatial Planning at the Norwegian University of Life Sciences. She is currently head of PhD education at the Department. She has worked as a practicing architect and as a planner in the town planning offices in Tromsø and Oslo. She has been a research fellow at the Norwegian Building Research Institute and a research fellow and research manager at the Norwegian Institute for Urban and Regional research. Inger-Lise Saglie has been a member of the boards for several research programmes in the Norwegian Research Council and abroad. Her research focuses on urban development with a particular focus on environmental issues.

Per Gunnar Røe is Associate Professor in the Department of Sociology and Human Geography at the University of Oslo. His fields of interest are urbanism and urban planning, suburbs and suburbia, spatial mobility and place, belonging and place making. His recent work includes editing a special issue on governance and change on the urban fringe in the journal *Urban Research and Practice*, as well as articles on suburbanism in the Oslo region. He is also a collaborator in the international research project Global Suburbanism.

Karen Lykke Syse is a cultural historian. She is Associate Professor at the Centre for Development and the Environment (SUM), University of Oslo. Her research interests pivot around environmental history and policy. Her focus is on environmental discourse and practice; agrarian and arboreal landscape studies; social and cultural aspects of hunting and angling, and nature in Early Modern thought. Recent publications include: *From Land Use to Landscape: A Cultural History of Conflict and Consensus in Argyll* (2009).

Kine Halvorsen Thoren is Professor at the Department of Landscape Architecture, Norwegian University of Life Sciences (UMB). She has written extensively on parks and the green structure.

Even Smith Wergeland received his MA in History of Art from the University of Bergen in May 2007. He has worked as an art mediator at the Rogland Museum of Fine Arts in Stavanger and as a lecturer in the Department of Linguistic, Literary and Aesthetic Studies at the University of Bergen. In 2008–2009, he was engaged as an advisor in the Department of City Planning in Stavanger municipality. Since February 2009, he has been employed as a PhD researcher at the Institute of Form, Theory and History at the Oslo School of Architecture and Design. The working title of his PhD thesis is 'From Utopia to Reality – the Motorway as a Work of Art'. The thesis, which deals with the aestheticization of motorways in post-war architectural theory, is written within the framework of the multidisciplinary research project Routes, Roads and Landscapes.

Introduction: Nature, Urbanism and Liveability

Mark Luccarelli and Per Gunnar Røe

Oslo, no less than other large cities, faces the difficulty of adapting to demands to reduce greenhouse gas emissions while securing green and public spaces and developing qualities of city living. As with many of the developed world's metropolitan areas, Oslo has made generous use of fossil fuel dependent transportation technologies to access ever larger hinterlands in search of space for populations who have demanded ever increasing standards of comfort as well as choice of where to live, work and shop. As the urban region faces the demand to decrease its fossil fuel dependency, existing modes of transportation and patterns of land use will need to be evaluated, which gives opportunity to raise the central question of green urbanism: the relation of urban design/urbanism to nature. We focus on a city, Oslo, because evaluating urban functions and qualities in light of ecology and ecological design should take place within a specific spatial, political and cultural context.

Oslo is an old town, but a relatively young city, and it has developed extensively in the 20th century as Norway has urbanized. While it is undeniable that Oslo has been marked by many of the same non-sustainable practices and outcomes that have characterized other cities in the Western world, it is our contention that Oslo also has a long history of being a 'green city' in the sense both of having a strong self-awareness of the city in its wider geographic setting and a long tradition of conservation of natural resources and green spaces. Significantly, 'green' in this sense is related to culture and to cultural memory – that is to practices, landscapes and monuments that provide an historical context for identification with place. This constitutes an important resource in building a local green urbanist discourse.

At the same time, as a systems approach to city and regional planning, green urbanism must confront the policy of densification, as explored in Parts II and III of the book. The effort to make the city denser, and thus more transit oriented, may threaten pockets of green areas and destroy monuments – a paradox for urban planners and policymakers. Our position is that this conflict can be managed more appropriately and democratically if a broad perspective of the city, its traditions, historical trajectory and possibilities for constructive change is developed.

A critical history of green Oslo can make an important contribution to understanding the opportunities that exist today for bringing the urban structure into a still finer relation to its geographical and natural surroundings by applying ecological principles to landscapes of varying scales and thereby deepening Oslo's 'green city' reputation. We also believe that the specific character of planning decisions of the present and future should be informed by an understanding of place that relies on various aspects and modes of analysis: the cultural geography, including imaginations, representations, mythologies and discourses; natural setting understood as green spaces, landscapes and region; and the built environment, infrastructures and architecture.

At the same time the particular experiences of Oslo may provide an important comparative perspective for urban regions of a similar size and type. The literature on medium-sized cities like Oslo is sparse compared with the amount of research on global and major cities and conurbations. In this respect, while Oslo is a city with its own characteristics, it bears resemblances as well as specific differences to similar large metropolitan areas – a fact which can be of great significance in looking at future trajectories of the city.

Backgrounds to Green Urbanism: The Sustainable City and the Commons

In 1987 the United Nations initiated the World Commission on Environment and Development, popularly termed 'the Brundtland commission'. The commission's book *Our Common Future* (World Commission on Environment and Development 1987) launched a broad movement towards sustainable policies within different sectors in society. Urban politics and planning was of course no exception, as the world was approaching the point in time where the total urban population would exceed the rural. The urban challenge got its specific chapter in the report. However, cities and urban regions had long ago been considered the locus of environmental problems; indeed they were often seen as the epitome of environmental degradation. The rapidly growing industrial cities of the 19th century, like Manchester and other shock cities, are cases in point. In fact, the first large-scale environmental urban policies and plans were developed because of a growing public awareness amongst the middle and upper classes who feared revolts as the consequence of horrible environmental conditions borne by the poor and working classes in such cities (Hall 2002). Even if much of the later environmental movement has been anti-urban, and has envisaged alternative settlement patterns (like 'eco-villages'), it is clear now that sustainability cannot be detached from urbanization and urbanism. Focusing on the urban world, the Brundtland commission stated that within the frame of sustainability and sustainable development, the city should develop in such a way as to reduce its impact on nature, and the environment, as well as limiting social environmental inequalities, thereby taking future generations into consideration (World Commission on Environment and Development 1987).

The Brundtland Commission report focused on cities in 'developing countries', meaning cities outside the industrialized Western world. Within the growing discourse on urban sustainability, however, increasing attention was given to the cities that had experienced de-industrialization, especially in the USA, Canada, Australia and Western Europe, because of their large share in energy use and emissions of pollutants, and because of the urban sprawl characterizing many of these cities and urban areas. The sprawling automobilized North American city, exemplified by cities like Phoenix, Arizona, has become the dominating worst-case representation and the prime target within the urban sustainability discourse. Sprawl has become the antithesis to sustainability, even recognized by observers sceptical to anti-sprawl policies, like Bruegmann who states that '…a key phrase in the latest round of anti-sprawl has been "sustainability"' (2005: 148). Indeed, one of the most important developments of the sustainability discourse has been conceptualization of 'the sustainable city'.

The implementation of sustainable city policies has the potential for reinvigorating environmental discourse. Arguably, environmentalism today has become so attuned to policies designed at the global scale that it has lost local initiative and consequently suffers from diminished public support (Dowie 1995, Meyer 2005). The *global turn* in environmentalism, specifically the increasing attention to global climate change, as well as the heavy emphases on international regulation of greenhouse gas emissions and technical fixes have narrowed green discourses. There is no contradiction, practically speaking, between global climate discourse and the sustainable city, of course: indeed, achieving a sustainable urban policy would be greatly aided by substantial agreements on global emissions of greenhouses gases. Nonetheless, global environmental discourse has had the effect of distracting attention from the local dimension of the global climate crisis and other systemic environmental problems. The sustainable city is a place-based policy strategy, which seeks to address the underlying causes of global climate change: over-consumption and hyper-mobility. This is very much a policy question; it must go beyond palliatives to a broader policy, but this policy should not be conceived as turning the clock back. As Dennis and Urry point out, 'You never change anything by fighting the existing reality. To change something, build a new model that makes the existing model obsolete'. A new system of transport that goes 'after the car', they continue, requires incorporating some of the indisputable advantages that characterizes personalized, flexible car travel while avoiding the high energy use of the existing car-centred 'system' (2009: 9). The sustainable city discourse alone cannot create a new model, but it has helped to rethink mobility and there has been recognition that the present spatial order has subverted important social values that might be recovered: the idea of the commons – the spaces and landscapes that residents share, the public square and the park – for one. To go beyond this should mean connecting with political discourses aimed at developing and empowering the public, as noted in the idea of the 'urban commons' and in the political philosophies of 'civic pragmatism' and 'communicative action' (Amin and Thrift 2002, Dewey 1926, Habermas 1991; Sorkin 1992, Minteer 2006). It would certainly mean linking planning discourse more profoundly to the 'cultural' sphere. If once we had 'master plans' that functioned as a blue prints, now we should have interdisciplinary projects to identify visions, discourses and potential outcomes – projects that should be ongoing. This volume represents the genesis of such a project. The change is, in part, a response to the diffused loci of authority in contemporary society. This is simply another way of saying that planning must reflect on and adapt to cultural processes which provide not only changed circumstances and conditions, but openings for reframing old proposals. To do this requires moving beyond sustainability to 'green urbanism'.

Green Urbanism

Alongside the concepts of urban sustainability and the commons, a range of discursive offspring have developed: 'Green Urbanism', 'Eco-cities', 'Urban ecology', 'Compact Cities', 'Resilient Cities' and so on. It is not our intention to evaluate the usefulness of these concepts for urban policies and planning or their discursive origins. In this book we have chosen to focus on Green Urbanism (Beatley 2005) for several reasons: firstly because it may contain a broad range of perspectives and policies, and thereby fits with the broad and interpretative approach in this book; secondly because it deals with urban design and planning and the relationship between humans and their environments; thirdly because the term 'urbanism' involves a social and anthropocentric perspective needed when reflecting on the social, cultural and political sides of sustainability; and fourthly because 'green' invites an exploration of how urbanism may be understood in relation to and as a reflection of the natural world.

The broad implications of green urbanism require exploring the relation between urban design/urbanism and nature. According to Timothy Beatley, cities that exemplify green urbanism strive to live within their ecological limits, to be green and function in ways analogous to nature, that is, to mimic natural functions by achieving a circular rather than linear metabolism, to move towards local and regional self-sufficiency by achieving a symbiosis with their hinterlands, to facilitate and encourage more sustainable, healthful lifestyles and to emphasize a high quality of life and create highly liveable communities (2000: 4–8). These three concepts – nature, urbanism, and liveability – and their interrelation are useful starting points for discussion of green urbanism.

Nature

Green urbanism builds on the sustainability discourse of the 1980s (Brundtland 1987) by elaborating how redevelopment of existing urban areas would promote environmental sustainability and green values. Sustainability discourse has conceptualized ecology through modernization, whereby the application of new technologies requiring investment generates economic activity in the process of meeting environmental goals. It was an answer to the old debates of the 1960s and 1970s which pitted economic interests, including manufacturers and unions, against environmental activists and lobbyists (Dowie 1995). This position has been important to sustainability discourse and green urbanism today, but it is nonetheless a limited use of the idea of ecology.

Green urbanism may be distinguished from ecological modernization in that it encompasses ecology as a core value (Beatley 2000: 5). In our view, green urbanism is not against ecological modernization, but sees it as inadequate and easily subverted: it can become merely another bureaucratic imperative subservient to established economic interests. What is required is a more pragmatic application of ecology that measures ends of urban design in terms of the means of ecological principles. One important context of ecological design is developed out of the human inclination toward place identity. We suggest green urbanism can work best when it develops a strong local identity. Consequently, the chapters in Part I of the volume first explore the roots of green urbanism in Oslo's historical development, especially regarding land use and housing, but also in terms of cultural attitudes and activities related to the environment. The other context for applying ecological principles is stated by Beatley in Chapter 1 of the present work: biophilia or the human attraction to all living beings – a wider natural world to which we are drawn by desire and necessity.

Nature may be advantageously understood as ecological function, operative as a regional ecosystem that must be preserved in order to provide the fundaments of life – air, water and waste disposal. But it is also a principle of design, which has implications for how the built environment might be arranged and rebuilt. The imperative of nature in green urbanism is to seek out the logic of ecosystems and find creative technologies and techniques that acknowledge those relations among living beings; this also requires finding appropriate images, designs and discourses that encourage parallel developments. If we are to overcome the reign of 'technological reason' (Marcuse 1964) as an environmental *raison d'être*, we need a way to talk about, think about and dream about urban designs that are rooted in natural ecosystems, and urban designs and systems that also mirror ecological design.

This is not to say that the city is simply a *part* of nature, because while we must live in ecological bounds, in so many ways cities reflect the social world and its priorities. The point is rather that cities could be seen as a *reflection* of nature and an expression, in part, of the attraction to nature, in some fundamental ways; and importantly that cities could mirror ecosystem design on whose viability all life depends. Many discourses in traditional planning and landscape architecture have equated 'greenness' with protected areas: green urbanism goes beyond this. It posits the question: in what ways does the city, or might the city, incorporate ecology in its design and function? Natural value and ecological design may be expressed in terms of urban landscapes.[1] There are three main contexts that should be considered: landscapes as reflections of ecological functions and

1 We are conceiving of landscapes here as variable and multiple, not the urban landscape as a whole.

principles; landscape as the organization of urban space/urbanism; and landscape as contributory to community. The application of ecological design to landscapes can occur at different scales listed here in order of size from large to small: the geophysical landscape, or 'local climate, plants, vistas, harbours, and ridgelands' that help define the 'special qualities of every memorable place' (Calthorpe 1993: 25); the ecological landscape, or the protected areas on which the natural ecological system that provides air and water to the city exists; the infrastructural landscape on which the city depends for its economic existence, and for its inhabitants' living; the designed urban landscape, or the green spaces, streets and public squares, architecture and public art; and the symbolic landscape of iconic structures that reflect an acknowledged sense of the city's identity. One important function of landscape design is to express what nature means to particular people in a particular place; this is a process that draws on a number of traditions and practices: outdoor or recreational culture; gardening; parkland preservation; naturalistic landscape design; green architecture; naturist design aesthetics.

Historically, Frederick Law Olmsted's regional landscape project around Boston, known as the 'emerald necklace', was important because it extended concern about the pictorial aspect of landscape to an ecological motive – preserving marshes, estuaries and forest in and around the city, and because it redefined what a park could become: turning it from a piece of land held in reserve to a part of a functioning regional ecosystem – today called a 'green structure'. Critically, Olmsted's conceptual movement from city park to the regional park opened the door to region-wide landscape preservation and appreciation by a wider public for both aesthetic and ecological purposes (Mumford 1938; Luccarelli 1995; Rybczynski 2000). This idea is reflected in contemporary bioregionalism that sees 'natural environment' as groundwork for all places and seeks to find ways to introduce natural values to urban situations; bioregionalism encompasses the new field of ecosystem management and bases it revisions of green spaces on the principle that nature is divisible into 'distinct communities of life' that may be brought even into the heart of the city through 'living corridors' (Thayer 2003: 33, 36, 188; Gordon 1990).The limitation of this approach has been a tendency toward preservation seen apart from urban design. Olmsted had already pushed landscape beyond pictorial representation by incorporating and representing natural processes; the challenge is to reflect natural designs in green spaces and streets through locally defined senses of space and attention to surfaces (Høyer 1999) and in architecture and the relation between buildings and urban spaces. Nature is also reflected in landscape understood as community (Olwig 2002). Indeed the idea of a communal landscape is a very old theme in the Western political tradition (Rousseau, Paine and Jefferson) and was also reflected in garden city thinking (Howard 2010/1898). As the celebration of nature is often a discourse about community, why shouldn't community be found reflected in understandings of nature? Don't parks, restored rivers, subways to the forest, river walks and displays of public art reflecting natural themes contribute to community and to a sense of the identity with a city?

Don't neighbourhoods reflect ecological patterns marked by variation, interaction and accretion much as natural ecosystems do?

(American) Urbanism: Historical and Discursive Contexts of the Green Turn

In geography, urbanism may be defined as the socio-spatial structures that delineate and organize cities. The complexity of the interaction between space and social organization defines the concerns of geographers. Human geographers see cities as complex entities functioning in the space between planning, design, policy and market dynamics. But green urbanism is based on the proposition that policy and design interventions are part of a larger and necessary cultural change in urban patterns of living that have evolved in an age of oil. The historical and discursive origins of this critique of modern urbanism and urban life tell us much about green urbanism's goals and challenges.

The most immediate origin of green urbanism is in the sustainable city discourse, which found a ready target in urban sprawl, particularly acute in North America. An early response to hyper 'automobility' was the compact city idea dating back to the early 1990s (Beatley 2005: 29). Sustainable city/compact city discourse has often been satisfied with promoting densification, which has been seen as more energy efficient. It's worth noting, however, that many of the new urban structures that are built in response to policy requirements to build in central locations, in places like Tjuvholmen and Skøyen in the Oslo area for example, are relatively energy intensive, both because of the activities taking place there and their architectural qualities: certainly, recent design trends to reappropriate the modernist glass box do not generally produce energy efficiency. These critiques while valid are besides the point for green urbanists; their chief concern is that densification will destroy the possibility of creating human-scaled, just and livable cities – a dream that lies at the heart of green urbanism and that has been expressed in various forms since the Italian Renaissance.

One manifestation of this wish for reshaping the modern city was the movement for designed decentralization. Reformist city planning in the United States in the 1920s and 1930s was associated with the garden city ideal and the RPAA;[2] they proposed garden cities as relatively decentralized, urban cores with a strong emphasis on residential or neighbourhood development. Subsequently, Jane Jacobs attacked these ideas, arguing that they helped reinforce suburbanism;[3] suburbia and the mania for highway building contributed to the undervaluing and destruction of inner city neighbourhoods, which represented the only places with sufficient density and centrality of location to support the dynamic qualities of urban life (Fishman 2000, Jacobs 1961, Whyte 1968). Jacobs was friendly toward

2 Regional Planning Association of America founded in 1926.

3 Which proceeded without the planned town centers the RPAA had proposed.

market solutions, however, and argued that American planning of the 1960s was too interventionist. What might she say today when market driven redevelopment schemes, often in collaboration with local or state authorities, threaten to remake city cores into over-developed beehive cities devoid of all green spaces?

Green urbanism has negotiated these histories of planning discourse; like the sustainability discourse on which it was built, green urbanism embraces the compact city idea, but at the same time it contests insensate densification and the language of technological functionalism (or 'ecological modernization') that dominates planning discourse today. Green urbanism should be understood as part of a larger tradition that has questioned the impact of speed, motion and massive scale on the modern city: it is aesthetic and social in its implications. The search has been for a formula for combining infrastructural and cultural questions into a unified approach. Lewis Mumford could be seen as contributing one of the key ideas of green urbanism with his container metaphor: cities may be seen as containers of social life he wrote (1961), which had burst at the seams because of excessive urban growth, casting the city's contents over a vast area in a rush toward urban diffusion. In a manner of speaking, green urbanism argues that the container must be mended, or more to the point, it must be reformed and reconstituted for both aesthetic appeal and ecological function. Indeed, green urbanism acknowledges this history and argues for a fundamental reworking of the idea of environment, as well as a contextualizing of that idea in terms of geographic structures.

The focus on the question of urban value or urbanity thus represents an important legacy that has contributed to green urbanism. Green urbanism then is not only a discourse aimed at strengthening urban structures, but an ideal of life that has surfaced in reform discourses over a period of several decades. Indeed, well before Jacobs, an ideal of urbanity may be traced back to reassessments of the city in the cultural criticism of Lewis Mumford, Richard Sennet, William Whyte and others in which urban spaces – squares, centres, cafés – are seen as inherent to human cultural development.[4] In their sense, 'culture' links aesthetic appreciation with social practices and public self-awareness.

Since the 1960s the city became a projection of the imagination of opposition, a search for alternatives and an expression of a longing for a better quality of life and for greater social cohesion. Urbanism developed as a cultural ideal, and as such reflects the post-1970 cultural turn in the humanities and social sciences. This in turn reflected the movement politics of that era that has responded to particular features of American life chief among them was anti-urbanism. The American middle classes had long exhibited a propensity toward hyper-mobility and careless use of natural resources which produced a reaction in the 1970s as evidenced

4 Jacobs presently occupies a central place in planning discourse, but Lewis Mumford's *The Culture of Cities* (1938) was an important work that predates Jacobs's by 25 years; despite her disapproval of his advocacy of garden cities and new towns, Mumford was a major of proponent of urbanity and was among the first to see the traditional European urban arrangements as a source for planning ideas.

by books such as *America the Raped: Who Needs Nature?* and *The Greening of America* (Moen 2002: 164). The counter-cultural hope that cities can produce a much larger share of their own needs for energy and food survives today in the idea of 'eco-cities' and, furthermore, is an essential part of green urbanism as evidenced in the analyses of 'urban ecocycles' and local renewable energy sources (Register 2006: 128, Beatley 2000). Urban activists of 1960s and 1970s saw the city as a new 'frontier' and their movement into cities as urban 'pioneering'; the eco-city concept, along with interest in architectural preservation, became the essential elements in the restoration of neglected urban neighbourhoods in New York and San Francisco – not to speak of Amsterdam and Copenhagen.[5]

The other source of the humanist urban ideal was the academic discussion galvanized by Jacobs, herself a 1960s activist, in praise of pedestrian spaces, urban communities and architectural-structural preservation. Urbanity may be measured by the walking city of the *flaneur*, which establishes the movement of the body and of the imagination through space as an important, indeed as an essential, though by no means the only, measure of urban scale and urbanism. Humanist urbanism may also be understood as a direct reaction to modern urban planning discourses, which held sway from the 1920s through the 1970s. Planning held forth for organizing space, or to be more precise, for its reorganization along the lines of techno-spatial modernization: urban spaces were defined primarily in functionalist terms which excluded both social and aesthetic questions (Boyer 1997/1983; Luccarelli 1995). The city had been conceived as a machine for production and rapid movement.

In Europe, modernist architecture argued for dismissing traditional conceptions of the pedestrian city and of mixed-use areas, and was in favour of the city of specialized spaces, mechanical movement, abstract geometries and vast scale as in Le Corbusier's 'city of the to-morrow' (1929/1971). By creating an aesthetic that matched these qualities, architects hoped to heal the divide between the technical and urban orders, between engineering and art, production and aesthetic value. It was a heroic attempt that soon collapsed (Jencks 1986/1973). In North America where modernism's influence was far less pronounced, unrestrained market forces magnified urban diffusion and disintegration, creating a virtual laboratory of discursive and cultural responses to the sprawling motor-and-highway dominated metropolis. The archetype of North American development had begun in the 1920s and produced one of the first anomalies of modern planning: the Los Angeles exhibit at the CIAM (*Congrès internationaux d'architecture moderne*) conference of 1927 held in Vienna. The scale of the map of the Los Angeles area dwarfed the European examples and was revealed as a metropolis that lacked the characteristics of a city: in particular it had no true centre (Fishman 1989). After the war, this 'fragmented metropolis' (Fogelsen 1993/1967) moved from anomaly to paradigm, becoming the recognized pattern of American urbanism. Already by the early 1960s, the geographer Gottman argued that vast highway networks

5 European cities were neglected during this period as well, but not as much, and variably.

produced enormous urban commuting zones that submerged boundaries within and among metropolitan areas; he called it 'megalopolis' (1961). Accelerating technological changes resulted in shifting production outside of centres, making the implosion of form inadequate to express a new regime of production, soon to be called 'technoburbia' or 'edge city' (Fishman 1989, Garreau 1992). Arguably what was being described was not down to the influence of a suburban ideal; it was a new form of urbanism, the consequence of diffusion of production made possible new forms of rapid communication and transport that ended the practice of vertical integration at one site (Fordism) and led to a 'disorganized capitalism' in which a post-Fordist production paradigm held sway (Lash and Urry 1987). The tendency was also present in Europe to a lesser degree.

In the American planning discourse of the 'City Useful' cities were places of commerce – not of living, of production not of reproduction. Failing to correspond to perceived values of nature, the city was devalued and became a mere instrument – a fossil of technical reason. By contrast, green discourse focused on preservation of natural habitats and reinvention of agrarian-linked ones. The first 'greening' of American cities – the advance of the pastoral suburb – arguably contributed to the development of diffused urbanism; it certainly dulled critical opposition. The creation of pastoral suburbs or new towns was understood as the solution to the 'problem' of urbanization, reproducing green visions of post-industrial age life (Gifford 1999: 36–9). This idea of greening as wilderness preservation and as de-urbanization may be understood as narratives of retreat from the city and return to nature that were often interpreted literally: unlike poetic pastorals of the imagination, 'sentimental pastoralism' left little place for the complementary theme of the return to civilization (Marx 1964). Suburbanization became an unintended consequence, one that unintentionally reinforced powerful technological and economic forces pushing urban sprawl and extensive diffusion of population in North American cities.[6] What makes the Scandinavian experience somewhat comparable to the North American are the cultural influences that helped determine the response to growing urbanism. It was a common inheritance of rural ideals and a reluctance to accept urbanization (White 1962, Facos 1998). The solution in both cases was to divide space into two spheres – one of life, one of work, and to conceive of the living environment outside of an increasingly functional city. The devaluing of urban environments, it has been argued, reflected 'green' (pastoral) literary and popular discourses that located ultimate value in agrarian, wilderness or exurban landscapes, which were seen as expressions of the natural world and which must be preserved as a bulwark against modernity (Cronon 1996).

Green urbanism reflects an advance against these disabling dualisms: the idea is that a balance between nature and culture can be worked out in a new kind of

6 The development of vast conurbations of slums in Africa and Latin America, and the growth of Asia mega-cities are competitors for the title of 'most extensive processes of urbanization' of our time.

urbanism.[7] There is also recognition that the crisis of the city is closely related to the crisis of environment. The response is to re-establish an urban order. Cities should be 'green' but they must also remain urban centres (Mumford 1961) with urbanist values.

Liveability: Dreams and Urban Spatial Structures

Although green urbanism is a cultural discourse reflecting the wish on the part of reformers for ecologically sustainable urban places, it is as much subject to the desires of urban residents as it is an expression of their ideals or cultural distinctions. Furthermore, urbanism as a motif must also respond to the geographic condition of urban systems, as they exist. Green urbanism surely affects the living conditions for city dwellers, and may therefore be seen as part of an urban social policy. But it may also be seen as a marketable asset, as we will discuss later.

Much effort has been made to create better living conditions in cities, and a better quality of life for its citizens. Modern urban planning, its visions and regulations, has focused on this. And in the 19th century, industrial city betterment was certainly needed, and could be operationalized to concrete measures of pollution, housing standards, sanitation and so on. However, and increasingly as living standards among most urbanites rose in Western cities, quality of life and living conditions were understood to have a cultural dimension. Of course, living conditions are not objective; neither are they a purely a subjective matter, but they are to some extent a result of intersubjectivity, culture and discourse. This is true when comparing cities, urban districts, neighbourhoods or groups of people.

How urban densities may influence living conditions and liveability may serve as an example and is certainly relevant for this volume. The perceived level of crowding (a negative representation of density) is not directly related to actual population density, but rather how housing areas are designed to secure privacy, fresh air and sun (Newman and Hogan 1981). In addition, there are of course cultural differences in space and time when it comes to living and coping with density: the difference between dense Asian cities like Hong Kong and shrinking and sprawling cities like Detroit, where decreasing density is related to decay, is large and surely has cultural connotations. Another example is city or population size. Urban visionaries and planners have sought to find the ideal city size, ranging from Ebenezer Howard's 32,000 inhabitants to Le Corbusier's three million in the 'city of to-morrow'. Different approaches to city growth and size comes from what kind of urban living is envisaged, and what aspects of urbanism are prioritized; economic activities and businesses or the living conditions of city dwellers.

Even if it may seem hard to define, liveability has become a buzzword for city policies today. It is in some ways the postmodern version of living standards,

---

7 This is not to imply an unqualified endorsement of the new urbanism movement in the United States.

focusing on opportunities, culture and leisure, in addition to ordinary measures of urban qualities, related to physical environment, housing and health. According to Pacione (2003) a wide range of social, economic and environmental needs must be satisfied in order to attain the goal of a liveable city – also those outside the regulative powers of urban geographers, planners and designers. Besides, the subjectivity of the objective environment must be acknowledged (Pacione 2003: 29). We would add that it is important to deconstruct the policies of liveability, to reveal its underlying discourses and powers of domination. To what extent is the quest for liveable green cities a commoditization to secure an economic definition of liveability, and to what degree would it imply mobilization and participation amongst different groups within the city? How does a representation of urban liveability, aiming at interurban competition, differ from an intra-urban conception of liveability, aiming at the living conditions and everyday life of city dwellers? Finally, if liveability increasingly takes the form of the commodity, will it be compatible with green urbanism?

The focus on liveability representing the quality of city life does amplify certain aspects, and especially those that can be operationalized and quantified for comparison between cities. Such comparisons are increasingly made by researchers as well as by popular magazines and reports. The rating of a city is believed to guide people and companies making decisions on where to travel, move and settle. As McCann (2004) points out, the rating of cities as good or bad places to live may have a significant effect on the ability to compete in the national and global market place for inward investment, industry, tourism and new residents. A good rating will aid and confirm urban entrepreneurialism and 'boosterism' strategies, and add to city pride. A bad rating may worsen city image, make it less attractive to the creative classes, investors and tourists, and even contribute to the downward spiralling of a city.

City politics and strategies aiming at competitive advantages and good ratings may advance some aspects of green urbanism, like designed parks and public transportation, while attributes not so easily marketed may be downplayed, like biological variety, greenhouse gas emissions and social segregation. Developing a green urbanism must therefore be based on a broad strategy taking different aspects of liveability into consideration.

Within the Norwegian political context, however, green urbanism, being a multifaceted and multilateral process, has not been a central issue. In the making of a policy for developing sustainable cities, and with the Brundtland report as a starting point, urban densification soon became the focus of state policies to avoid further urban sprawl and automobile use. However, this was not the only possible model for future, and greener, urbanism. In addition to the densification approach or compact city model that considered vehicular emissions from the city as a whole, a second model focused on the ecology within the city, its green structures and biodiversity, and produced a vision not compatible with large-scale densification. A third model is the regional approach, a decentralized concentration of settlements, and a sort of compromise between the other two models.

Green Oslo?

Oslo has a green reputation and has won two prizes for such from the European Union for developing as a green city. Partly, of course, this is a matter of city self-promotion, place marketing or 'boosterism'. It is a response to a new global economic order in which cities have had to compete in the marketplace for investment and prestige. At the same time, the concern with the city as an expression of civic pride and cultural identity is a discourse as old as the Renaissance, indeed even older: its origins are in ancient Greece and it periodically characterized political discourse in post-medieval Europe (Hohenberg and Lees 1995).

As a result of the unquestionable strong presence of nature, in its landscape and representations, Oslo may be seen as having a favourable position in the quest towards green urbanism. The presence of nature and the rural may also be supported by an argument that Oslo has a short history as a major city, weak urban traditions and a weak urban intellectual base for the promotion of urbanity as a value for itself. However, a strong anti-urban discourse does not lead to a green urbanism, as Frank Lloyd Wright's vision for a 'broadacre city', part of the American search for pastoral environments, proves. Rather it could support the argument that cities oppose nature, or that cities should be dissolved (or spread out), causing increased transport and thereby an unsustainable development. Besides, the representation of Oslo as a green city, or the pervasiveness of nature in its culture, may overshadow the 'realities' of how green Oslo is, when all aspects of green urbanism are taken into consideration.

The reason for us focusing on Oslo is not that the city is a 'best case' or that its policies and plans suggest a 'best practice' to be promoted across the world. Neither is it the purpose of this book to decide if Oslo is representative within a group of cities. And we are not aiming at giving an accurate assessment rating Oslo amongst other cities with regard to 'greenness', by using indicators for urban sustainability or the widely used methodology measuring ecological footprints. Rather, we wish to ask how Oslo may be understood and interpreted as 'green' from different angles and within different theoretical perspectives. This includes both subjective and intersubjective experiences and practices within the city, images and representations of the city, as well as policies and plans for the city. In this way we hope to problematize the representation of Oslo as a green city, but at the same time to envisage how the green urbanisms of Oslo are constructed.

This said, the case of Oslo is, in our view, of interest for an international audience. The unquestionable greenness of the city is not merely good luck, or because the result of the city's location in a sparsely populated country, but also a consequence of political decisions and plans. It may also stem from a culture responsive to at least some aspects of green urbanism. The traditional 'outdoor culture' (MacKaye 1928) or *friluftsliv* in Norwegian, at least insofar as it symbolizes and enacts the national identity and pathos, could be seen as a cultural capacity to develop green practices and green urbanism, or at least it may be said to produce a cultural inclination favouring such practices.[8] Influential actors, as well as the public, may wish to protect nature in the city and its outskirts, catering for outdoor activities like hiking, jogging, skiing and so on. But, as mentioned, it may also be of hindrance, because there doesn't seem to be the need for a greener city, and because people may think their practices are green enough.

A discourse necessary to deal with these paradoxes of the public domain seems to be developing. Media debates deal with the issue, comparing Oslo with other European cities regarding, for example, public transport and urban green spaces. However, a better understanding of the social and cultural complexities of 'green Oslo' is needed. How do people experience and interpret Oslo as a green city? How does this fit with major representations of green Oslo? And what kind of urbanism has been developed as part of different discourses on Oslo? How are policies, plans and urban structures developed as a result of this? We believe these questions not only are of importance to understand the relationships between culture and urbanism in Oslo, but they also give a better understanding of the hindrances for developing a green Oslo. These issues will also be of importance when reflecting on green urbanism and when resilient strategies are to be reached in other cities; also because the literature often points in the direction of European and Scandinavian cities to find cases of green urbanism (Beatley 2000, Newman, Beatley and Boyer 2009 and more).

At present, Oslo, like most thriving metro centres, enjoys and also suffers from a concentration of economic activity in the city centre and a burgeoning population. Green urbanism offers the following insight: over-concentration of economic and cultural activity in the main centres is almost as bad a consequence of current economic imperatives as was the diffusion of economic and cultural activity during the era of mass (auto)mobility which reached its epitome in North America and virtually destroyed many urban centres there. Green urbanism asks: is there not a middle ground? Could the development of new infrastructure around mass transit and energy distribution be employed to encourage multi-nodal metropolitan centres?

In Norway in the 1990s, densification was seen as a means of achieving a compact urban form, and a way of implementing the 'vague notion of the compact city model' (Saglie 1998: 16). This policy was also pursued in other European countries, and within the EU. While this strategy in some countries included the construction of new, and compact, cities, the focus in Norway has been on existing cities, and the potential to increase densities within already built-up areas (Saglie 1998: 82). Despite its intuitive and indubitable effect on transportation (Newman and Kenworthy 1989; Næss 2006, 2009), the densification strategy

8 The collective value of the practice of outdoor culture may be undermined as, or if, it reverts to an (more or less consciously) understood marker denoting distinction and distance to other social classes and socio-cultural strata.

and the compact city ideal have been critiqued. There are mainly two types of arguments: one that deals with the limitation in expected transport reduction, and a second argument that deals with the societal consequences of densification, including the possible development on green lands of value to the city population. Breheny (1995) argues in line with the first argument that little is to be gained through densification, and with regard to the second that the political and social consequences of redevelopment are too big. Some even argue that increased densification, and overcrowded cities, may increase the flight to suburbs and beyond (Blowers and Pain 1999). The consequences will certainly depend on how densification is achieved, both regarding process as well as physical planning and design, and the actual densities: the densities of 19th-century Lower East Side in New York City, for example, are certainly not desirable.

Therefore more nuanced approaches have been needed to handle the environmental, social and cultural complexities involved, and green urbanism may be a starting point. Nonetheless, the richness of policies, measures and examples characterizing the emerging green urbanism discourse needs to be elaborated and supplemented with knowledge, local and contextual, of how green urbanism is constructed or obstructed through culture and discourse. Innovations, we often forget, actually arise from indigenous conditions and resources. In *Cities and Civilization*, Peter Hall argues that the idea of the city, and the achievement we call the city, arises in a consciousness of itself which can be the consequence of a 'long accumulation of psychological and social traits, a … cultural reproduction over generations, among people living together … that makes them especially amenable to art or to thought' (1998: 20–21). The accumulated capacity to define life through ideas and art is not new: it defines the Italian renaissance for example.

Approaches, policies and plans dealing with these complexities also challenge research based on simplified notions of human action, which attribute human motivation to instrumental and utilitarian motives. These sorts of explanations do not deal with how people interpret and handle the changes involved, and therefore may at best be of limited use. This is part of the motivation for this book.

Green Urbanism as Asset: Privatization of Public Value

Green urbanism may not only be pursued for the sake of the world and to reduce global environmental problems, like carbon emissions. Other non-environmental issues and interests are also at stake. Green urbanism may also be turned into a political strategy to satisfy political and economic goals. Indeed Harvey states that it is impossible to separate the cultural politics of place from social processes underlying the development of place (1996: 320). The greening of cities and city green spaces may thus be utilized in three ways: (1) as an asset for a neighbourhood or urban district; (2) as an intra-urban asset, for the city population as a whole; or (3) as an interurban asset, in competition with other cities for capital and people.

In the first case, greening of the city may enhance the quality of life for people in certain neighbourhoods, where for example green spaces are developed or new public transport lines constructed. However, such local improvements may also influence the housing market and contribute to gentrification, that is, the gradual replacement of poorer citizens with middle-class or upper-middle-class residents. Urban sustainability programs frequently include measures that focus on the environmental and economic components of sustainability, while the social dimension of sustainability remains underrepresented, according to Pearsall (2010), who has studied the negative impacts of Brownfield Redevelopment strategy in New York City for elderly residents, renters and residents receiving government assistance. If cities are experiencing increasing social polarization and spatial segregation, such changes or green improvements may in sum contribute to a socially divided city also with respect to environmental qualities (Sassen 2006). Green lifestyles and green urbanism with parks, up-to-date public transport systems and energy efficient and health promoting housing may be developed in the high-end of residential areas, contributing to an increase in housing prices, while little is invested in green infrastructures to support green lifestyles in poor neighbourhoods.

In the second case, the greening of a city may lead to enhanced liveability, measured for the population as a whole. Measures and ranking of cities according to liveability indicators has become big production, reported in newspapers, magazines and so on. The yearly ranking by the lifestyle magazine *Monocle* is one of these presumably influential assessments of city liveability: in 2010 Munich, Germany, was ranked as the most liveable city, while Oslo finished 18th. Another example is the annual ranking conducted by the *Economist.* The liveability indexes focus on some aspects of green urbanism, such as public transportation, environmental issues and access to nature, in addition to social and economic factors.

In the third case, green urbanism may be an asset in the global competition between cities and their countries. Cities are in the forefront of globalization as command points, creative hot spots, cultural melting pots and as sites for research and development and advanced production. Of increasing importance for city politics is marketing and promotion, focusing on urban qualities for living and business establishments. Green urbanism, or its representations, may therefore be transformed into strategy to sell the city on a global market, in competition with other cities for people and capital. Urban sustainability or green urbanism may not only be part of urban governance towards increased urban liveability of public value, but may as well be interwoven with competitive entrepreneurial policies to make cities more attractive for business investments and a highly educated creative class. Keil (2005) has written extensively about the interrelationship between neo-liberal policies and nature.

The cultural distinction of green urbanism may thereby appear at the level of subjective and intersubjective level (urban green lifestyles), at a neighbourhood level (housing areas or urban districts represented as 'green') and at a city level (green cities). Through social practices and consumption of artefacts, housing and travel, some people and places stand out as green, and may contribute to the culturalization and stylization of green urbanism. The paradox is consequently that the consumption of green may be environmentally unsustainable, or that the greening of cities is socially unsustainable.

If not a 'best case' for green urbanism, Oslo may be a 'special case', reflecting its location on the edge of European urbanization, as Helle (2006) points out. Studying Oslo may be of comparative value therefore within the context of the development of 'ideal types' which would be based on a typology drawn from a set of characteristics. Oslo may be classified in regard to established criteria, such as the general type of urbanism (European), the regional culture (Scandinavian), the level of social provision (articulated social welfare state) as well as the particularities of the local/national traditions in perceiving landscape and cities. This may not represent a complete set, but it certainly lays out some of the more relevant characteristics which must be considered green urban research and planning. Oslo as an ideal type, then, differs markedly from an example of an American 'green city', which will reflect the differences both in spatial and social structure between North America and Europe, among other factors. Yin (2003) used Jane Jacobs's study of New York City, its sidewalks, the role of its parks and so on as an example of how a single case study may be of use in developing an ideal type. We are suggesting that a series of studies could create a typology of green urban ideal types that could spur successful planning efforts by inviting comparison and contrast and by forcing planners and stakeholders to reflect on the real conditions which they face, as well as the assets they command. This book is not a strictly analytic case study, but an exploration of Oslo's green spaces, landscapes and imaginations, and its politics and plans.

Plan of the Book and Chapter Summary

This book gathers and draws into relation the perspectives, discourses and modes of analysis necessary to an urban planning practice that responds to the challenges of the environmental crisis. We are living through a time of fragmentation of the cultural fabric, deteriorating national institutions and weak global governance which has produced, among other things, multiple perspectives and agendas on the part of stakeholders. This has arguably helped democratize planning, but it has also left us vulnerable to market-driven priorities. We have been left with a crisis of governance. The arrival of a new planning discourse for the 21st century, whether created by a discursive turn or emerging as a new hegemonic paradigm, should combine social and natural ecologies to ecological modernization, taking into consideration the specifics of place and the local traditions of place awareness, as well as global social and environmental concerns. It must articulate a culture- and place-specific understanding of 'green' while articulating a critical discourse analysis and developing a coalition of stakeholders capable of making institutional changes.

This book, *Green Oslo*, points toward a discursive formation for green city planning and implementation. In our view to succeed, such green urban discourse should consist of three elements which correspond to the three sections of this book: (1) conceptions of environment and visions of change; (2) complications presented by academic discourses and historical examples of planning; and (3) the necessary planning trajectories and policies required to implement green urbanism.

Part I

In the opening segment, four chapters consider the various means by which 'green' may be understood in relation to urban life. The alternatives presented here may be read as contributory to the development of a successful green urbanism in Oslo and by extension in other cities. That process of development begins with a vision of green cities that becomes specific to a locale on the basis of knowledge of the urban landscape, its history and cultural development.

Timothy Beatley is an influential theorist of green urbanism. In Chapter 1, he argues that environmental reform requires cities that permit and encourage 'close contact with nature'. Fundamentally, human evolution has been a coevolution with our surroundings: the environmental crisis indicates that the process of coevolution is endangered. Restoring the health of the environment is necessary to the survival of the human species and as cities become the predominant human environment we must think of them in biological terms. Citing the work of biologist E.O. Wilson, Beatley suggests that healthy cities are 'biophilic'; that is, they are cities that develop an effective environmental ethic based on a strong understanding of and feeling for nature, infrastructural developments that permit access to green areas and policies that limit the impact of resource use and preserve local (and global) biodiversity. He concludes that Oslo ranks high in these characteristics and may be considered one of the world's most biophilic cities, though much remains to be done: he outlines additional steps that might be undertaken to forward the project.

Beatley's argument is that a green city is 'natural', that is to say it harbours wild plants and animals in green spaces large and small. This idea of nature, then, is less a unified landscape than a set of functioning green spaces established amidst and within the city. In Chapter 2, Karen Lykke Syse discusses the history and significance of the largest of the green spaces in the Oslo area (much of it within the municipality of Oslo): the impressive forests (or *Marka*) that flank the city to its north. Syse treats the *Marka* as a cultural landscape, which has had significance in the life of the city over the last 300 years as Oslo grew from a small manufacturing city to the Norwegian capital. Certainly the forest is a vast natural area that provided and continues to provide an outlet for Oslo residents to experience the natural world, but Syse emphasizes the *Marka* as a rural environment, which grew up next to the city and was utilized for economic purposes as much as it was valued as an extension of the natural world. The development of the landscape arose out of sustainable practices of environing (or situating human activity in natural surroundings). What Syse shows is that a natural landscape became part of the cultural fabric, a set of connections that weld it to the city, which it helps define. Landscape in her sense rests on experience of both use and appreciation over time.

This phenomenological landscape is also the theme presented in a very different way by Fredrik Chr. Brøgger in Chapter 3. Brøgger describes how Frogner Park, which lies near his boyhood home, became nature for him. The centrepiece of the park is a highly formal sculpture garden (built 1929–42) containing the works of Gustav Vigeland that reflected formalized European, modernist and national ideals represented in a highly symbolic landscape. The park the young Brøgger came to appreciate was the old Frogner Park designed earlier by Marius Røhne, a gardener and landscape architect influenced by Frederick Law Olmsted. Brøgger thus demonstrates something of the struggle in Oslo to maintain a preference for natural environments, and he also shows that biophilia reflects the environmental imagination: the natural motifs of Røhne's design both reflected natural forms and interpreted them through design motifs. The resulting biophilic landscape both enabled biodiversity and provided an enchanted refuge for play.

The forest preserve surrounding the city and the parks within it are fundamental aspects of green urbanism, but they are reserves – green spaces that have been set aside from residential or commercial development. In Chapter 4, Brown and Luccarelli query the possibility of 'green' design within residential areas by re-examining an old experiment: the Ullevål garden city ('Ullevål Hageby'). The landscape of the garden city with its housing patterns, architecture, street layout, public areas and boundary demarcations is the focus of the article. Proceeding by considering the natural and social implications of garden city design, both at Ullevål and in famous English examples, Brown and Luccarelli work out a series of design principles that show how picturesque, wild and regional interpretations of nature are imbedded in garden city design. The authors are less interested in whether the garden city model could be revived at the edge of today's cities, emphasizing instead how these design principles that might be applied to the redevelopment of existing urban landscapes.

Part II

The five chapters that comprise Part II examine green urbanism from various discursive perspectives and planning positions. Clearly green urbanism is more than a vision: it is has something to say to planning discourse, architectural discourse, housing discourse, among others. At the same time, the histories and concerns of these discourses have much to say to the idea of green urbanism. This part of the book presents some of the problematics and possibilities of green urbanism in various discursive and spatial contexts.

In Chapter 5, Peter Hemmersam points out that green urbanism and similar concept terms, for example sustainable cities, green cities, eco-cities and so on, have become advertising language by private interests and public authorities interested in marketing cities within a global economic environment of interurban competition. Furthermore, by analysing the city's official architectural policy (the PBE 2009), Hemmersam points out that the real agenda is a highly technical regime of energy savings and overlooks other aspects of green urbanism: issues of ecology, social equity and cultural preference. Hemmersam's critique extends to what he sees as facile conceptions of densification, which he argues do not necessarily produce green results.

In Chapter 6, Guttu examines the ideas behind the formation of residential green areas, through Oslo's modern history. He explains the major shifts in planning for residential areas with ideological changes, influencing planning discourse and practices. Importantly, he argues that there seldom is a direct link between ideology and planning, because planning discourse is moulded of different ideologies, and because there is a time lag between ideas and implementation.

In Chapter 7, Røe examines the paradoxes of green urbanism within the spatial and representational context of suburbs and suburbia. Suburbs and suburbia have been considered antithetical to green urbanism, because of their contribution to urban sprawl, encroachment of nature and unsustainable lifestyles. However, suburbs are increasingly differentiated, in form and social composition, making generalized approaches less useful. The author investigates specifically how green spatiality, or the planning and practices concerning green recreational spaces, have, in different ways and forms, been part of very different suburban developments during the modern era. In a case study of a large suburban housing project from the early 1970s, he exemplifies the dilemmas of density versus green spatiality, and top-down discourse versus a participatory and flexible approach to design.

In Chapter 8, Wergeland deals with the dilemma of green urbanism by investigating the role of driving spaces, that is, urban roads and motorways. He argues that the green agenda fails to deal with certain aspects of the landscape, like the unruly and dangerous places in a city, as driving spaces are. In a case study of a major urban road in Oslo, he reveals the troubled relationship to the idea of the green city.

Chapter 9 asks us to consider how existing landscapes are actually treated. Landscape architects Hedfors and Florgård eschew recent trends to raise a more fundamental issue: the loss of an ethic of concern and care for the landscapes, which we already have. Raising the banner of the great landscape gardener of the last century, Ian McHarg, and stressing the need for 'maintenance', they argue that only a commitment to sustaining 'living landscape' and to designing with nature, as McHarg puts it, can create the conditions from which green urbanism may develop.

Part III

In the last part of the book, the planning itself and its outcomes are investigated. This is done through analysis of the history of planning in Oslo, and through case studies of planning projects. The aim is to give an overview over green aspects in planning, and exemplify how green concerns have been dealt with.

Grundt presents a broad environmental assessment of Oslo, today, as a starting point in Chapter 10. Then he gives a historical overview, stretching from the founding of Oslo in 1000 AD to the politics of today, of the perceived environmental challenges and policies to deal with them. His conclusion is that the municipal efforts to reduce the environmental impacts have been fairly successful. The weak points are the protection of green areas, biodiversity and the ever present problems of pollution of air, water and ground. Grundt ends his chapter by reflecting on the troublesome urban politics with regard to green or ecological strategies.

In Chapter 11, Jørgensen and Thorén make a review of overall planning of Oslo from the early 1900s and to the present time, examining how parks and green spaces have been part of master plans in this period. The authors argue that the period can be divided in two: first, a growth period where, paradoxically, the planning of green spaces and structures had its prime time; and second, a decline period, where economic difficulties partly can explain the observed fragmentation of Oslo's green structure.

Based on recent research, Falleth and Saglie in Chapter 12, the last chapter of this book, focus on the compact city ideal, and how it is balanced with the protection of green spaces and the provision of a green infrastructure. In the professional discourse the issue of green infrastructure does not seem to be downplayed, but rather it seems to be missing. This may give developers more opportunities to build with high densities, in line with the compact city ideal, but also according to economic profit. It is local civil society actors who seem to challenge policies and planning for the compact city through bottom-up protests against development projects, as the authors exemplify in two case studies.

Bibliography

Amin, A. and Thrift, N. 2002. *Cities: Reimagining the Urban*. Blackwell: Oxford.

Beatley, T. 2000. *Green Urbanism*. Washington, DC: Island Press.

Blowers, A. and Pain, K. 1999. The Unsustainable City?, in *Unruly Cities? Order/Disorder*, edited by S. Pile, C. Brook and G. Mooney. London: Routledge.

Boyer, M.C. 1997 (1983). *Dreaming the Rational City*. Cambridge, MA: MIT.

Buell, L. 2005. *The Future of Environmental Criticism*. Oxford: Blackwell.

Breheny, M. 1995. The Compact City and Transport Energy Consumption. *Transactions of British Geographers*, 20, 81–101.

Bruegmann, R. 2005. *Sprawl: A Compact History*. Chicago, IL: The University of Chicago Press.

Calthorpe, P. 1993. *The Next American Metropolis*. New York: Princeton University Press.

Corner, J. 1999. 'Eidetic Operations and New Landscapes', in *Recovering Landscape*, edited by J. Corner. New York: Princeton Architectural Press.

Cronon, W. 1996. The Trouble with Wilderness; or, Getting Back to the Wrong Nature, in *Uncommon Ground*, edited by W. Cronon. New York: W.W. Norton.
Dennis, K. and Urry, J. 2009. *After the Car*. Cambridge: Polity Press.
Dewey, J. 1927. *The Public and Its Problems*. New York: H. Holt.
Dowie, M. 1995. *Losing Ground*. Cambridge, MA: MIT Press.
Duany, A. and Plater-Zyberk, E. 2001. *Suburban Nation*. New York: North Point Press.
Ehrenfeld, D. 1981. *The Arrogance of Humanism*. New York: Oxford University Press.
Facos, M. 1998. *Nationalism and the Nordic Imagination*. Berkeley, CA: University of California Press.
Fishman, R. 1989. *Bourgeois Utopias*. New York: Basic Books.
Fishman, R. 2000. *The American Planning Tradition*. Washington, DC: Woodrow Wilson International Centre.
Fogelsen, R. 1993 (1967). *The Fragmented Metropolis*. Berkeley, CA: University of California Press.
Garreau, J. 1992. *Edge City*. New York: Anchor Books.
Gifford, T. 1999. *Pastoral*. London: Routledge.
Glotfelty, C. and Fromm, H. 1996. *The Ecocriticism Reader*. Athens, GA: University of Georgia Press.
Gottmann, J. 1961. *Megalopolis*. Cambridge, MA: MIT Press.
Gordon, D. 1990. *Green Cities*. Montreal: Black Rose Books.
Habermas, J. 1991 (1962). *The Structural Transformation of the Public Sphere*, translated by T. Burger and F. Lawrence. Cambridge, MA: MIT Press.
Hall, P. 1998. *Cities and Civilization*. London: Trafalgar Square.
Hall, P. 2002. *Cities of Tomorrow*. Third edition. London: Blackwell.
Harvey, D. 1996. *Justice, Nature and the Geography of Difference*. Oxford: Blackwell Publishing.
Helle, K, et al. 2006. *Norsk byhistorie gjennom 1300 år* (1300 Years of Norwegian Urban History). Oslo: Pax Forlag.
Hohenberg, P.M. and Lees, L.H. 1995. *The Making of Urban Europe, 1000–1994*. Cambridge, MA: Harvard University Press.
Howard, E. 2010 (1898). *To-morrow: A Peaceful Path to Real Reform*. Cambridge: Cambridge University Press.
Høyer, S.A.B. 1999. 'Things Take Time and Things Take Things: The Danish Landscape', in *Recovering Landscape*, edited by J. Corner. New York: Princeton Architectural Press.
Jencks, C. 1986 (1973). *Modern Movements in Architecture*. Hammondsworth: Penguin Books.
Jacobs, J. 1961. *The Death and Life of Great American Cities*. New York: Vintage Books.
Keil, R. 2005. Progress Report – Urban Political Ecology. *Urban Geography*, 26, 640–51.

Lash, S. and Urry, J. 1987. *The End of Organized Capitalism*. Cambridge: Polity Press.

Le Corbusier. (1929/1971). *The City of To-morrow and its Planning, translated from the 8th French ed. of Urbanisme by Frederick Etchells*. Cambridge, MA: MIT Press.

Luccarelli, M. 1995. *Lewis Mumford and the Ecological Region*. New York: Guilford Press.

Marx, L. 1964. *The Machine in the Garden*. New York: Oxford University Press.

Minteer, B. 2006. *The Landscape of Reform*. Cambridge, MA: MIT Press.

Marcuse, H. 1964. *One-Dimensional Man*. Boston, MA: Beacon Press.

McCann, E.J. 2004. 'Best Places': Interurban Competition, Quality of Life and Popular Media Discourse. *Urban Studies*, 41, 1909–29.

MacKaye, B. 1928. *The New Exploration*. New York: Harcourt, Brace & World.

Meyer, J.M. 2005. Does Environmentalism Have a Future? *Dissent* (Spring Issue).

Moen, O. 2002. 'Mobility, Geographic and Social: The American Dream and American Realities', in *Nation on the Move*, edited by C.A. van Minnen and S.L. Hilton. Amsterdam: VU Press.

Mumford, L. 1938. *The Culture of Cities*. New York: Harcourt, Brace & World.

Mumford, L. 1961. *The City in History*. New York: Harcourt, Brace & World.

Newman, P., Beatley, T. and Boyer, H. 2009. *Resilient Cities*. Washington DC: Island Press.

Newman, P. and Hogan, T. 1981. 'A review of urban density models: Toward a resolution of the conflict between populace and planner'. *Human Ecology*, 9(3), 269–303.

Newman, P.W.G. and Kenworthy, J.R. 1989. *Cities and Automobile Dependence: An International Sourcebook.* Aldershot: Gower.

Næss, P. 2006. *Urban Structure Matters: Residential Location, Car Dependence and Travel Behaviour*. New York: Routledge.

Næss, P. 2009. Residential Self-Selection and Appropriate Control Variables in Land Use-Travel Studies. *Transport Reviews*, 29, 293–324.

Olwig, K. 2002. *Landscape and the Body Politic*. Madison, WI: University of Wisconsin Press.

Pacione, M. 2003. Urban Environmental Quality and Human Wellbeing – a Social Geographical Perspective. *Landscape and Urban Planning*, 65, 19–30.

Pearsall, H. 2010. 'From Brown to Green? Assessing Social Vulnerability to Environmental Gentrification in New York City'. *Environment and Planning C: Government and Policy*, 28, 872–86.

Register, R. 2006. *Ecocities*. Gabriola Island, British Columbia: New Society Publishers.

Rybczynski, W. 2000. *A Clearing in the Distance*. New York: Scribner.

Saglie, I-L. 1998. *Density and Town Planning: Implementing a Densification Policy*, NIBR's PLUSS-series, 4–98, Oslo: Norwegian Institute for Urban and Regional Research.

Sassen, S. 2006. *Cities in a World Economy*. Thousand Oaks, CA: Pine Forge Press.

Sorkin, M. (ed.) 1992. *Variations on a Theme Park.* New York: Hill & Wang.

Thayer, R.L. Jr. 2003. *Lifeplace: Bioregional Thought and Practice.* Berkeley, CA: University of California Press.

Treib, M. 1999. 'Nature Recalled', in *Recovering Landscape*, edited by J. Corner. New York: Princeton Architectural Press.

Waldheim, C. 2006. 'Landscape as Urbanism', in *The Landscape Urbanism Reader*, edited by C. Waldheim. New York: Princeton Architectural Press.

White, M. and White, L. 1962. *The Intellectual versus the City*. Cambridge, MA: Harvard University Press.

World Commission on Environment and Development. 1987. *Our Common Future.*

Whyte, W.F. 1968. *The Last Landscape*. Philadelphia, PA: The University of Pennsylvania Press.

Yin, R.K. 2003. *Case Study Research.* London: Routledge.

PART I
Visions, metaphors and patterns

Chapter 1
Biophilic Oslo

Timothy Beatley

Biophilic Cities: What Are They?

To be happy, healthy and productive, urbanities require daily contact with nature. To what extent can cities and urbanized regions be said to be natural, and to provide the basic conditions for urban citizens to live a life in close contact with nature? This is the subject addressed in this chapter, with a specific goal of evaluating whether, and the extent to which, Oslo can be said to be a *biophilic city*.

Harvard biologist and myrmecologist E.O. Wilson popularized the concept of *biophilia*. By this, he meant that humans have coevolved with nature, that we carry with us an ancient brain that needs and is deeply referential of nature and other forms of life. More specifically, Wilson describes biophilia as 'the innately emotional affiliation of human beings to other living organisms. Innate means hereditary, and hence part of ultimate human nature' (Wilson 1993: 31; see also Wilson 1984, 2007). To Wilson, biophilia is really a 'complex of learning rules' developed over thousands of years of evolution and human-environment interaction.

That we need nature in our lives is abundantly clear and the empirical evidence continues to mount. We are happiest, emotionally (and physically) healthiest, and most productive when working and living in close proximity to nature. Urban buildings that are green and natural create the conditions for more productive work, and in the case of hospitals and health care facilities, the conditions for maximal healing. Schools that are awash in natural daylight are environments in which students excel and test scores demonstrate this benefit (for example, see Kellert, Heerwegen and Mador 2008).

The evidence also shows that there are many ways in which urban environments can provide that access to nature, and many different forms that nature can take. It is having access to parks and natural areas close by to visit and use, and trails to visit and walk along, but it is also the views of nature that we sometimes do not even pay attention to but that in innumerable ways provide visual access to nature.

The nature in cities is both big and small and varied in quality. It might take the form of a small neighbourhood park, a green rooftop or vegetative wall or a large block of forested land at a city's edge. It is also in the form of the many animals that occupy urban spaces, from birds and bats to the invertebrates in urban soil and to larger species like wolves and coyotes that frequent spaces at the edges of cities and towns. It remains an open question as to the mix of different kinds of nature, and nature experiences, that urbanites need, though clearly a diversity of opportunities is desirable.

Elsewhere, I have written at great length about what constitutes a biophilic city (Beatley 2011a). Table 1.1 lists some of the more important dimensions of biophilic urbanism or biophilic cities and some possible indicators or metrics, some but not all readily available for the city of Oslo. It is important to note from this table that the physical quality or attribute of a city, how green or natural the city is, is but one dimension. If residents care little about this nature, and fail to make the time in busy daily lives to visit and enjoy it, it is difficult to argue that such a place is a biophilic city. It requires more physical or design qualities, though these are certainly important.

Table 1.1 Some important dimensions of biophilic cities (and some possible indicators thereof)

Biophilic Conditions and Infrastructure – Percentage of population within a few hundred feet or meters of a park or green space; – Percentage of city land area covered by trees or other vegetation; – Number of green design features (e.g. green rooftops, green walls, rain gardens); – Extent of natural images, shapes, forms employed in architecture, and seen in the city; – Extent of flora and fauna (e.g. species) found within the city.
Biophilic Behaviours, Patterns, Practices, Lifestyles – Average portion of the day spent outside; – Visitation rates for city parks; – Percent of trips made by walking; – Extent of membership and participation in local nature clubs and organizations.
Biophilic Attitudes and Knowledge – Percent of residents who express care and concern for nature; – Percent of residents who can identify common species of flora and fauna.
Biophilic Institutions and Governance – Priority given to nature conservation by local government; percent of municipal budget dedicated to biophilic programs; – Existence of design and planning regulations that promote biophilic conditions (e.g. mandatory green rooftop requirement, bird-friendly building design guidelines); – Presence and importance of institutions, from aquaria to natural history museums, that promote education and awareness of nature; – Number/extent of educational programs in local schools aimed at teaching about nature; – Number of nature organizations and clubs in the city, from advocacy to social groups.

Urbanists and city planners have special opportunities and unique obligations to advance biophilic city design, utilizing a variety of strategies and tools applied on a number of geographical and administrative scales. The agenda is one that must extend beyond conventional urban parks, and beyond building-centric green design. It is about redefining the very essence of cities as places of wildness and of restorative nature, from rooftops to roadways to riverfronts. It is about understanding cities as places that already harbour much nature and places that can become, through bold vision and persistent practice, even greener and richer in the nature they contain.

As Table 1.1 summarizes, then, a biophilic city is certainly partly defined by the qualities and biodiversity present and designed-into urban life, but also the many activities and lifestyle choices and patterns, the many opportunities residents have to learn about and be engaged directly in nature, and the local institutions and commitments expressed, for instance, in local government budgets and policies. What a biophilic city is or could be is an open question. As a tentative starting point, I offer some of the following as key qualities of biophilic cities:

- Biophilic cities are cities of abundant nature in close proximity to large numbers of urbanites; biophilic cities are biodiverse cities that value, protect and actively restore this biodiversity; biophilic cities are green and growing (biologically, that is) cities, organic and *natureful.*
- In biophilic cities, residents feel a deep affinity with the unique flora, fauna and fungi found there, and with the climate, topography and other special qualities of place and environment that serve to define the urban home; in biophilic cities, citizens can easily recognize common species of trees, flowers, insects and birds (and in turn care deeply about them).
- Biophilic cities are cities that provide abundant opportunities to be outside and to enjoy nature through strolling, hiking, bicycling, exploring; biophilic cities nudge us to spend more time amongst the trees, birds and sunlight.
- Biophilic cities are rich multisensory environments where the sounds of nature (and other sensory experiences) are as appreciated as much as the visual or ocular experience; biophilic cities celebrate natural forms, shapes and materials.
- Biophilic cities place importance on education about nature and biodiversity, and on providing many and varied opportunities to learn about and directly experience nature; in biophilic cities there are many opportunities to join with others in learning about, enjoying, deeply connecting with and helping to steward nature, whether through a nature club, organized hikes, camping in city parks or volunteering for nature restoration projects.
- Biophilic cities invest in the social and physical infrastructure that helps to bring urbanites to a closer connection and understanding of nature, whether through natural history museums, wildlife centres, school-based nature initiatives or parks and recreation programmes and projects, among many others.

- Biophilic cities are globally responsible cities that recognize the importance of actions to limit the impact of resource use on nature and biodiversity beyond their urban borders; biophilic cities take steps to actively support the conservation of global nature.

There are undoubtedly other important qualities or dimensions not listed here, but this is a useful starting point in assessing Oslo's biophilic *bona fides*. Whether Oslo is a biophilic city will depend on how well it reflects these different dimensions or aspects of biophilic urbanism. My ability to fully assess the city against the full list of criteria above has been limited by distance, data and language, so conclusions here are tentative and preliminary. They are based on a combination of site visits and walks in the city, a review of relevant and available data and a review and reading of key city documents and plans (such as the city's proposed Green Plan). While findings are preliminary, it is nevertheless clear that, in many important ways, Oslo is highly biophilic, with potential to further reinforce and strengthen these biophilic and natural elements.

Oslo might contain much nature close by, reflecting biophilic sensibilities at home, but through its large ecological footprint and long-distance supply lines it may be indifferent to impacts on nature and other forms of life that occur beyond the city's boundaries.

Is Oslo a Biophilic City?

Even the most unobservant visitor to Oslo will likely comment on its green and natural qualities, and seems that nature is never far from where one is even in the centre of this compact city. On almost every physical measure, residents of Oslo have access to an extraordinarily high degree of nature. In visiting Oslo, nature seems all around and one is never very far away from it.

Oslo is a relatively small city: 454 km^2 with a population of about 570,000 (part of a larger metropolitan area of around 1.2 million residents; Institute of Transport Economics 2009). Oslo's application to become a green capital of Europe summarizes the impressive access to green areas provided to residents, making the claim that 94.3 per cent of the city's residents live within 300 metres of a green area (in 2006). Another metric by which to judge a city's greenness are green spaces and parks *per capita*. Again, Oslo's number is very high: 47m^2 (Oslo City Council 2007: 23). In Oslo, there is abundant nature and it seems most residents are not very far from a park or green space (Oslo Kommune 2008).

The trends may be slightly more concerning as the latest numbers suggest a small decline in access to parks and green spaces. While 94.3 per cent is very high indeed, this is down from 94.6 per cent in 2001 and for access to parcels over a hectare in size, the decline has been from 89.1 per cent in 2001 to 88.8 per cent in 2006 (Oslo Kommune 2009: 18). Some small amount of loss of green space in the city is the likely culprit, something that jives with the concerns of some local environmentalists who feel deference and priority has been given to building projects over parks (for example, including the planned construction of the new US Embassy, which will result in the loss of some open land). This may also be the result of reurbanization of inner city locations (a positive trend from a sustainability perspective), where access to green areas is lower. The measure of green space *per capita* has also shown some decline (down from $49m^2$ in 1998), but this decline is largely a result of population growth. Nevertheless, the larger pattern and elements of extensive green spaces remain intact, but these slight trends do suggest that the city must be ever more diligent (and creative perhaps) in protecting the nature that does exist.

And the green and biophilic qualities of the city vary depending on which part of the city is considered. Access to green spaces is highest, as would be expected, in the outer areas of the city, and lowest in the inner city. Providing green features and green elements in the densest and oldest parts of Oslo remains a challenge. And as the city's master plan anticipates further densification (about two-thirds of the projected growth in housing by 2025 is slated to occur in the central areas of the city; See Oslo City Council 2007: 56–9), there will need to be careful thinking about how urban biophilia in these areas of Oslo can be advanced at the same time.

Oslo has prepared an impressive Green Plan, or green structure plan (*Grontplan for Oslo*), which carefully and methodically takes stock of these green qualities and also lays out an ambitious future vision for further restoring and growing nature in the city. The existence of this plan is itself an indicator of a biophilic city (more on the vision below), but it also provides a useful set of analyses of the extent to which Oslo residents have access to nature (Oslo Kommune 2007). The Green Plan argues for the importance of not a single kind of park, but for three sizes of parks, each providing all important green access to urbanities while providing for different types of outdoor experience and preserving necessary ecological functions.

There is an impressive network of trails and a considerable ability for residents to reach these green areas in one or more ways, including walking, bicycling and transit. The Oslo environment provides an unusually high degree of connectivity for residents, and many ways to get around without an automobile. The trail system is quite impressive. The extent of existing and planned trails is some 365 kilometres, and judged on a *per capita* basis, one of the most extensive urban trail efforts anywhere (for example, comparable to that of Anchorage, Alaska).

The city's efforts at getting residents out of their cars is commendable, and an important dimension of a biophilic city. The city has further invested in its bicycle infrastructure, for instance, adding some 1,000 new bicycle parking spaces in the last decade, as well as a city bicycle fleet. At the same time, it has been gradually reducing parking spaces in the city centre and restricting parking and cars in other ways. And the city has set its sights on becoming even more of a bicycle city, adopting the goal that by 2015 some 12 per cent of trips in the city will be made by bicycle (Oslo City Council 2007: 51). That said, bicycle mobility remains underdeveloped and underemphasized in Oslo compared with many other European cities: by comparison, in the city of Copenhagen nearly 40 per cent of home to work trips there are made on bicycle, and there has been an extensive effort to extend, expand and enhance the bicycle network and infrastructure, including intersection lighting for bicycles, synchronizing lights to speed bicycle travel along key corridors, and initiating Green Cycle Routes to provide commuting through and alongside green areas (see Beatley 2011 for a review of these initiatives)). A 2009 University of Oslo study puts the number of bike paths in Oslo at one-third of what exits in Copenhagen (122 km within the municipal boundary; Institute of Transport Economics 2009: 24). Interestingly, a recent report of the Norwegian Institute of Transport Economics shows the considerable health benefits (and monetary value) of increases in bicycle ridership in several Norwegian cities (Norwegian Institute of Transport Economics 2010).

Biophilic cities are cities that provide residents with viable alternatives to cars, and Oslo does this admirably. Already about half of the journeys of five kilometres or less are made on foot, and there have been solid improvements in the city's public transport system including extending the city's metro system. It has been estimated that some 88 per cent of the city's residents are within 300 metres of a transit stop or station with a service frequency of one hour or less. Oslo famously installed a toll ring road for cars, one of the first cities to impose a form of congestion charge. This sends a signal, and has had the effect of reducing car traffic in the city, but has also raised a considerable income: 45 per cent of the revenues from it are specifically designated for use in improving public transit.

The city's longstanding commitment to compact urban form is impressive and has done much to both promote more walkable, sustainable urban conditions, and to create conditions in which a large portion of the city has close proximity to the forest edge. The success of these efforts at urban densification is clear. Between 2000 and 2009, the city's density, as a result of these planning policies, has been estimated to increase about 11 per cent, a quite considerable increase in a relatively short period of time (from 37.9 persons per hectare to 42.3; Institute of Transportation Economics 2009: i).

Compactness in turn reduces the amount of land needed to accommodate urban growth and in many ways makes possible the conservation of large blocks of forests at the city's edge. Indeed, reportedly two-thirds of the municipality's land area, a little over 300 km^2, is either in forest or lake, and so an immense amount of nature is owned and managed directly by the city of Oslo and not far from residents (Oslo Kommune 2008). Within the urbanized areas, about 20 per cent is considered green and open land, so there is a fairly sharp contrast between the urbanized city and the large forests that surround the city to the north and east. And beyond these municipal forests lie even larger expanses of nature.

The city's large area of forests, or *Marka*, provides, moreover, diverse and abundant recreational opportunities. There are some 70 bathing lakes, some 100 rivers and lakes suitable for fishing and 460 kilometres of cross-country ski trails. This is made possible in large part because of the impressive spatial planning principles by which the city is essentially surrounded or encircled by extensive forest land. And the delineation of the forest boundary line or border (so-called

Markagrensa) is a significant step, indicating that those areas beyond the line (beyond the urbanized city) are to be protected and not developed.

Oslo's current municipal master plan makes a strong endorsement of sustainability. Goal 5 of the plan states that 'the city of Oslo will follow a sustainable path of development' and the plan lays out a series of strategies for advancing this goal, including compact urban development (Oslo City Council 2008: 46). The plan states clearly that new growth 'is to take place in the city centre, near the fjord, near stations and other public transport nodes, and along important public transport axes, where a mixture of commercial and industrial activity, cultural institutions and services and housing will strengthen the multifunctional city and create vibrant urban environments' (Oslo City Council 2008: 48). The city will further densify, but its restriction on building height (of 42 metres) will be held firm. The plan of the future is one of strengthening its compact urban form, hand-in-hand with strong protections for the city's forests and green areas. 'The boundary between the forest and building zone will be maintained and fjord, green spaces and rivers and streams will be protected' (Oslo City Council 2008: 48).

The city's application to become the European Green Capital succinctly summarized commitment to compact urban form: 'The overall strategy of preventing urban sprawl has been Oslo's policy for many years. This strategy primarily includes densification around collective transport networks and nodes as well as improvement of the public transport system. In addition, building in the greenbelt around the city is strictly prohibited' (Oslo City Council undated: 3).

Oslo's spatial plans, and new draft Green Plan, place nature even more front and centre in this city. And the citizens of Oslo appear regularly to visit and enjoy this abundant nature. Nature is not just near and seen from windows, but actively visited and enjoyed. The forest or *Marka*, especially, holds a special cultural and social importance, and is a place visited frequently by residents of Oslo. By one statistic, 81 per cent of the city's population had visited the forests around the city during the previous year (Oslo City undated).

Undoubtedly, other lifestyle indicators or metrics would similarly show an outside or outdoor orientation that fosters biophilic connection. The proximity to the fjord and the maritime traditions suggest an importance to sailing and boating, and a relatively high percentage of Oslo residents own their own boats. Through improvements in the city's waste-water treatment system, the waters of the inner fjord have seen dramatic improvement in recent years, and now are fishable and swimmable, though it is not clear how much of these activities occur. A coastal or port city has special biophilic opportunities related to water and aquatic habitats and Oslo appears well positioned to further strengthen these aquatic bonds.

While residents of Oslo have amazing access to green space and nature, there are some biophilic experiences that may be less available in the city. There appear few allotment gardens available for residents to grow some of their own food, a common (and important) feature in other European green cities (for example, Copenhagen, London, Freiburg). As concerns about the food security and resilience of cities and urban populations increase, there may be many other reasons to support food growing in urban areas, and some cities (for example, London) have even developed food strategies. Putting one's hands in soil, tilling and harvesting vegetables and flowers is an ancient activity that many enjoy and that should increasingly be part of any biophilic city. In some cities, efforts have been underway to stimulate food growing and food production even on the rooftops and balconies of high-rise buildings (for example, Montreal) and Oslo may wish to follow suit.

The ability to get residents out of their cars, and onto bicycles or on foot, is a major challenge for any biophilic city in the industrialized north, and here Oslo excels as well. Car dependence is the norm, and has the effect typically of sealing us off from the natural world and from each other. Oslo has many cars to be sure, but also has taken a number of steps to moderate their impacts, and the numbers suggest good results in creating conditions for a more 'walking' city. One reported statistic is that some 85 per cent of the city's children get to school either by walking, bicycling or by public transit (Oslo Green Capital undated). Compared to typical American cities, then, there is relatively little use of cars to shuttle kids to and from school, perhaps setting them down a life path of walking or bicycling, and offering at least some of the conditions of a more biophilic urban childhood.

Planning for green elements in Oslo appears also to recognize the many other important benefits provided by this 'green infrastructure'. The green plan speaks of the importance of green spaces, vegetation and forests in addressing climate change and in moderating its impacts: 'At a time when increased rainfall is expected and a warmer climate in our part of the world, use of vegetation and green areas as a tool to achieve better local climate, local air quality and water balance will be even more important … Green areas that are very important for the local climate, either because they constitute ventilation ducts [conduits? or corridors?], fresh air supply areas, wind protection areas … should be preserved and strengthened' (Oslo Kommune 2007).

Biophilia is, I would argue, multisensory as well, and as much a function of the sounds, smells, tastes and bodily sensations one feels in living environments. We often privilege the visual and ocular, and sometimes forget how essential these other senses are to our connections to place and environment, and how the senses work together to deliver the sensations and experiences that make up the ordinary nature around us. And it is often the totality of these sensory inputs that determines our experience of place (the breeze on our skin, the sounds, the smells, the tastes of the food we eat), along with the sights.

Sound is an especially important element in this biophilic sensory mix that is often undervalued or overlooked. The sounds of nature (birds, frogs, insects, swaying trees in the wind) are unique to particular places and represent unique place qualities. Sounds (and smells) often trigger important memories and can help to bind us to places.

I am not aware of a sound map or sound study conducted for Oslo (such a step would be of considerable value), but, as in many contemporary cities, noise is a concern. The Oslo Noise Action Plan attempts to at least partially take this on through the designation of quiet areas in the city (some 14 of these places have been identified where noise levels are below 50 dB: 'Areas which can offer recreation, outdoor experiences, and/or cultural activities in surroundings sheltered from or distant from dominant noise sources'; Oslo Noise Action Plan, quoted in Oslo City, undated: 38). The city's river corridors would comprise about half of these (linear) quiet zones. About a quarter of the city's population suffers from an excess of noise (over 55dB during the day), a not insignificant amount, and it is mostly connected to roads and auto traffic. The more the city is able to discourage driving and put in place other steps to control car noise (design quieter road surfaces, discourage studded tyres in winter, for instance), the greater will be the possibilities that residents will connect with the more biophilic sounds all around them that they presently find difficult to hear.

That is a good step, and partly responds to an EU directive, but more could be done to understand the positive soundscape, natural and human made; not just the avoiding the negative or noxious, but actively rekindling a sense of the sounds in the city and region that are enjoyed, that evoke deep recollections and memories and that are immensely enjoyed by those who hear them.

Quiet areas may be parks but could extend to other places in the city and offer the possibility that by reducing the impact of car noise, the more biophilic and therapeutic sounds might become more audible to residents. There are, moreover, a number of points in the city where distinctive natural sounds can be heard, including the fast water and roar of the waterfalls along the Akerselva, suggesting again other soundscape benefits of fully restoring the city's eight streams on their way to the fjord.

Knowledge about the nature around one is a significant prerequisite for appreciating it and stewarding over it. And our knowledge of native species of flora, fauna and fungi is notoriously limited. We are apt to recognize many things, but the native butterfly or bird is often not one of them. Similarly, we rarely recognize or understand the sounds of the animals around us, for instance the sounds of owls or night birds, commonly heard but not very consciously listened to.

A significant difficulty for nurturing a biophilic urban citizenry is the sense of profound disconnection such populations feel and experience from the nature around them. And much of that nature, especially the aquatic or underwater variety, strikes us as especially remote, visually, psychologically, and ontologically separate from the terrestrial world of *Homo sapiens*. Oslo's proximity to, and historical dependence on, the sea (the fjord and the Baltic and North Seas beyond) makes this an especially important aspect of biophilia to consider.

Globally, this sea-city disconnect exists despite the fact that our increasingly urbanized world is perched on the edge of vast seas, and that all souls living on Earth live, as oceanographer and underwater explorer Sylvia Earle reminds us, on the Blue Planet (see Earl 2009; see also Beatley 2011b). Urbanites know scarce little about the magical and wondrous nature a few feet below the surface. To test the visual knowledge of prominent and important oceanic predators, those at the top of the ocean food chain, I recently gave the students in my large introduction to urban planning classes a visual test – asking them to identify several species of especially distinctive, indeed even iconic, ocean creatures. These included: a sperm whale, a Northern Right Whale (known as the urban whale along the Atlantic coast of the US), a leatherback sea turtle and an Atlantic Bluefin Tuna. The results are a bit discouraging and show a rather limited knowledge about even our largest sea-life, those at the top of the food chain. The sperm whale is one of the world's most visually distinctive whales, with its square front and teeth rather than baleen. Nonetheless, only a relatively small number of student respondents (33 out of 173) could correctly identify it. Another image was of a leatherback sea turtle and results were similarly worrisome: while a majority recognized it as a sea turtle, only a single student, out of some 180 or so, correctly recognized it as a leatherback turtle, a surprising result given this sea turtle's very distinctive and unusual appearance. The news was not all bad, and I was pleasantly surprised at how many of the students could recognize a tuna (66 out of 181). Still, the overall result reinforces just how remote sea-life and ocean environments are from us, even when the distance from our comfortable terrestrial world may be not be particularly great; the perceptual, psychological distance is immense and how to bridge these distances and forge greater awareness of and concern by the urban population remains a major challenge for an aspiring biophilic city.

Oslo has much going for it in this regard and perhaps a real chance to kindle an ocean sensibility that other more landlocked cities may find difficult to match. And there is already considerable evidence that the near-sea waters anyway have gained in attention and priority (more about this below).

Oslo has already achieved much in the area of urban sustainability. In a recent Siemens-funded study of some 30 major European cities, and the development of a European Green City Index, Oslo ranked third overall (Oslo Kommune 2008). It was in fact the best performing city on CO_2 emissions, largely a result of the large amount of hydro-electric power in the country, but also a result of its expanding district heating system and improvements in public transit. The city's *per capita* CO2 emissions rate is quite low (2.19 tonnes per person), as is its *per capita* energy consumption (about 95 gigajoules). But here the city continues to make strides, including a new municipal sewage treatment plant extracting biogas that could power as many as 250 municipal buses.

The city has also adopted an energy action plan aimed at enhancing energy efficiency for building and expanding the percentage of local energy coming from renewable sources. And the city has set some ambitious targets, including reducing greenhouse gas emissions from public buildings by 95 per cent by 2030. The city has also embraced environmental management systems in major ways, eco-certifying all of its municipal agencies and offices.

Biophilic Growth and Development

Visiting Oslo, I was impressed by the new thinking there about how and in what ways the city's waterfront growth might take advantage of the unique and special water setting. Indeed, urban designers and planners in the city are speaking of Oslo as the 'fjord city', and imagining how future building will connect with and respect this special aquatic environment. A fjord city plan, laying out principles and visions for the future of the waterfront there has been prepared, with some interesting concepts: a new fjord city park, new 'activation' of the water areas (for example, creating spaces for boating and swimming), a harbour-wide promenade and limits to the height of new buildings in the city to ensure visual connections with the water, among others (City of Oslo undated). 'The waterfront will be a part of the greater common urban space, which everyone can use and acquire a positive connection to. The water is a magical, organic and universal element and Oslo's motto: "The Blue and the Green, the City in Between" can now be given enhanced and significantly clearer contents. 'Through the construction of an attractive, unbroken harbour promenade and a set of public spaces, a beautiful coastline is achieved and the fjord is interconnected with the neighbouring city districts' (City of Oslo undated: 2).

These new waterfront developments have not been without their controversies and critics. Some in Oslo still believe the buildings will be too high, and there are worries about the increase in the value of land and housing in these areas (and as a result, a worry about how socially and economically inclusive these new areas will in the end be). There are some 40 islands in the Oslo fjord and they are mostly accessible to the public. Oslo provides exceptional public access to its coastal and shoreline environments, and suffers not from the high degree of privatization and cordoning off from public use that commonly occurs in many other parts of the world.

The recent and dramatic design of the Oslo Opera House exemplifies some of the possibility here. A design by the architectural firm Snøhetta was chosen in a design competition, with a primary design element being enhancement of connections with the harbour and inner fjord. 'The project is designed in such a way that the fjord is not blocked off by a new building. In fact, the opera site is generously given back in the form of an extended urban space that allows for new experiences of the larger landscape of the inner Oslo fjord' (Opera Forlag undated, p.46). The building's dramatic, almost four-acre, granite roof slopes into the fjord water, creating a large and unique urban plaza, and allowing visitors to virtually touch and dip into the surrounding aquatic world.

The connection to the water is to be further strengthened with a planned waterfront promenade, and a larger, longer coastal trail is under development. Moreover, in Oslo, public access to the shorefront is well protected, compared with other coastal nations where private property and development essentially create walls, making public access difficult or impossible.

Indeed, Oslo has already had considerable success and some international acclaim for its waterfront development. The district Aker Brygge, with its mix of office, shops and relatively dense housing, and very compact and walkable design, puts many residents in close proximity to the water and has created a destination for visitors as well. This complex, built during the later 1980s and early 1990s, is highly regarded by urbanists and frequently complimented. Jan Gehl, pedestrian guru and urbanist, gives it impressive accolades in his new book *People and Cities*, describing it an 'outstanding urban area' and noting the exceptionally sensitive combination of density, uses and design that make this district enticing. 'Careful consideration was given to density, mix of functions and good city space. Despite a high building density … the buildings do not seem tall, because those along the streets have fewer floors than those set further back … City space is well proportioned with active ground floor frontages, and thanks in great part to the good design, the area has become one of the few new urban areas in Europe where people actually enjoy spending time. Density is high, but it is the right kind of density' (Gehl 2010: 69).

And Aker Brygge has a number of other biophilic aspects, in addition to the important one of creating the context and conditions to entice residents to be outside and to walk and stroll. Much of the public art that one sees here reflects biophilic shapes and forms.

The city's new draft Green Plan seeks to further strengthen the city and region's green infrastructure, especially the major rivers that run through the city. The plan sets out the bold goal indeed of bringing all eight of these rivers fully back to the surfaces and restoring them as green and ecological features in the city. Figure 1.1 shows the main summary or aspiration map of the Oslo green plan, showing the eight

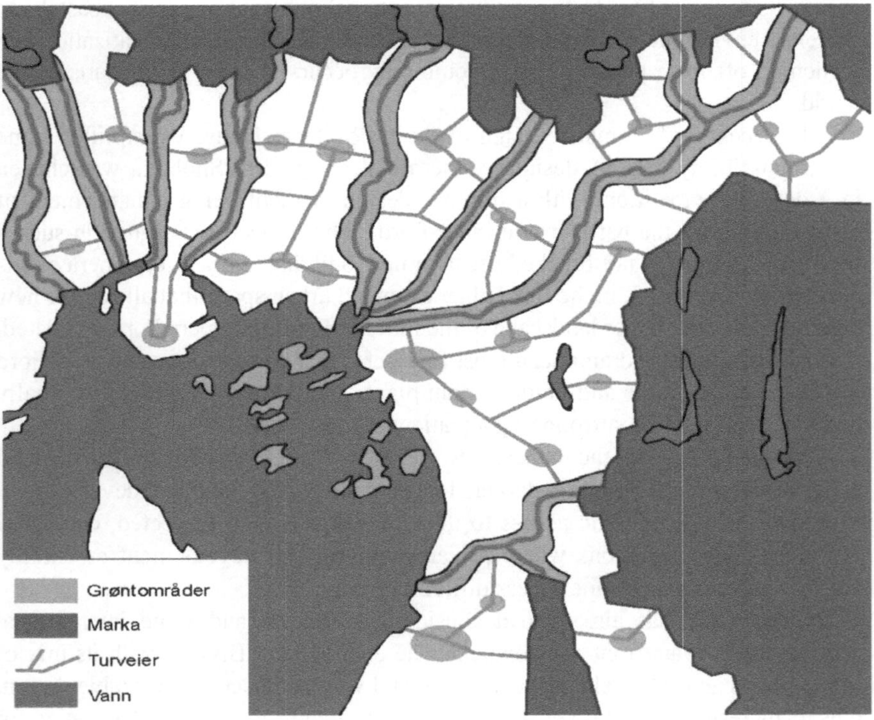

Figure 1.1 The Green Plan's vision for the future for Oslo: 'the blue and the green and the city in between'

rivers connecting the *Marka* or forest to the north and east with the fjord waters, as well as connecting a network of smaller green spaces and parks in the city.

The value and potential of restoring and bringing back to the surface its main rivers can be seen in one of the rivers, which is already in this more natural condition. An impressive green ribbon that runs through the heart of the city, the *Akerselva* (*elva* means river in Norwegian), was in medieval times the main source of water, then later where much of the city's industry was located. Today it is an active, popular place to cycle, walk and picnic. The river is at times quiet and calm, but in many places is fast moving and loud, and there are some unexpected and dramatic waterfalls. The trail and pathway system along the river is in many places on both sides of the river and the river is itself quite wide in parts, with small islands in the centre. It has hard surface pathways, elegantly trimmed with lines of stone. There are a number of footbridges along the route, providing opportunities to cross over and experience the river from the other side. While the river is crossed at a number of points by roads for cars, there are pedestrian underpasses that ensure continuity of the path system. There is a remarkable amount of development and housing along the river, large apartment blocks with ready pedestrian access to the river, which is one the main reasons this green element works, as it is not remote, but highly accessible to lots of residents, with the result being lots of walkers, joggers and bicyclists, and a feeling of safety even during the evening hours. There are new cultural amenities here along the river as well – new clubs, theatres, artists' studios – that help to further enhance its popularity and attractiveness, and that further generate foot traffic and visitation.

Another secret relates to the creative ways in which parks and open spaces are woven into the river corridor, places that on a sunny Sunday were full of people picnicking and barbequing. Others were dog walking and several were fly-fishing. There are lots of built-in benches and small spaces to sit and watch the river or read a book. The sounds of the river can be heard many metres away and are actually deafening near some of the falls. The banks sloping down to the river have been cut only along the edges leaving spectacular floral displays of yellows and greens on the day of my June visit. There is a municipal cutting policy evident here that mows only the perimeters, enough to show the spaces are being taken care of, while beautiful fields of flowers are spared. In strolling around, several times I encountered people who had clearly been collecting bunches of flowers to take home or bring with them to whatever social engagement they were on their way to.

Oslo might also be judged on its biophilic aspirations. It has prepared an impressive Green Plan that set some ambitious green goals and green targets for the future. The green plan sets goals and series of strategies for achieving those goals. As already mentioned, the first strategy is to develop a 'continuous green network of public green areas'," where a resident can walk or ride their bike from one of these parks and green areas to another. The Green Plan evaluates the access of residents and urban neighbourhoods to three sizes of parks: small, medium and large, and identifies spatial and access deficiencies in different parts of the city. And while the results are generally good, there are suggestions for ways that

Figure 1.2 The *Akerselva* is an impressive restored river corridor offering residents opportunities to hike, bike and picnic in the midst of trees, wildflowers and, in several spots, significant waterfalls

Source: Photograph by the author, 2010.

access to all three forms and sizes of parks can be assured. The city's green plan calls for 69 new parks in places that help correct the deficiencies in access.

Other Ways Oslo Might be Biophilic?

A biophilic city is defined in part by its sensibilities towards nature and its priorities, expressed in many different ways. Are buildings designed to reduce impacts on birds, for instance, or have efforts been made to design roads and infrastructure to minimize mortality for wildlife (many cities have invested in wildlife eco-bridges, underpasses and overpasses that permit safer wildlife movement). What steps does a city take to protect and care for its wild flora and fauna (or for that matter, its domestic pets)?

An additional measure of a biophilic city is the extent to which natural shapes, forms and images make their way into the design of buildings and urban design elements in the city. This is a difficult thing to judge without a more comprehensive effort at taking stock of built environments. Yet, even a cursory meandering uncovers a number of examples of this design motif, such as the maritime biotic shapes at *Aker Brygge*, for instance, and the many textures and shapes in the city centre or *sentrum*.

Non-human forms can and do make their way into the historical stories and mythology of a place. During a recent visit to Lisbon, I discovered that here animals, namely birds, figure prominently in the story of the patron saint of the city, St Vincent. More specifically, legend holds that after the martyrdom of this saint in the 3rd century, ravens (or crows) guarded over the body until it was buried. The crows have now become the city's mascot, and are even visible on the city's flag, and are prominently perched on the statute of St Vincent who occupies a prominent spot in downtown Lisbon.

The city of Oslo appears to be committed to effectively stewarding over the nature under its control. It appears to guard and keep stridently off-limits the *Marka*, just beyond the urbanized city. The city manages this land under ISO 14,001, and specifically employs a 'Living Forest' management standard and certification standard that appears to protect ecological functions, at the same time as allowing some amount of sustainable harvest. The Norwegian Living Forest management certification system sets out a series of standards, including requirements for leaving uncut a certain number of trees (an average of 5–10 living trees per hectare) for buffer zones in riparian habitats, and restrictions on altering dead wood and damaging terrain (Sverdrup-Thygeson, Borg and Bergsaker 2008). This is further evidence of the long-term stewardship values of the city.

How it seeks to manage or moderate its impact on nature beyond its municipal boundaries, another measure of its biophilic commitments, is unclear. The city's ecological footprint is large, and as with most northern and western cities, where personal consumption is high, there are significant ways in which the city imposes negative externalities on others. A 2002 study by the University of Oslo's Centre for Development and Environment calculates Oslo's footprint to be about 7.7 hectares per person, and requiring a land area of about 90 times the city itself (Centre for Development and Environment 2002). While smaller than other cities, this footprint is about three times the 'globally available biologically productive land': 'If the world's population had a similar footprint', the study notes, 'we would need *three* extra *planets* to live on' (Centre for Development and Environment 2002: 21). About half the footprint is due to food, which adds further support for findings ways to support urban agriculture and community food production in and around Oslo.

In a brief visit, and with limited data, it is difficult also to gauge the extent to which Oslo supports a biophilic lifestyle, and the extent to which residents care about and actively seek out nature. Some evidence of this was presented above, specifically the high percentage of residents who visit the *Marka* during the year. But, of course, this may be a low threshold, and more important perhaps is the number of people (or the percentage of the city's population) that visit the *Marka*, or another green space in the city, on a daily basis. To what extent is being outside a priority to Oslo residents, and how truly biophilic are the life patterns and lifestyles of the residents of this city? In the US, an astoundingly high percentage of the average day is spent inside buildings, or riding in an automobile; more than 90 per cent in fact. I must assume the percentage in Oslo is lower, but by how much?

The extent of outdoor- or nature-orientation might also be assessed by such proxies as the number of residents who own items such as bicycles and boats. I am unclear about the former, but it has been mentioned by several that owning and using a boat (sailboats and motorboats) in Oslo is very common, and this may in fact be one very positive sign that proximity to a fjord is taken advantage of and integrated into their lifestyles in ways it may not be in other coastal cities. One wonders about the planning and policy levers that might support or hinder boating and sailing. Are there sufficient docks and marinas, are biophilic considerations taken into account in how boats and sailboats are taxed (for example, ways to encourage boating) and to what extent is it easy and affordable to learn to boat or sail? These are important questions in a city like Oslo that has such immense aquatic resources.

How engaged Oslo residents are in watching nature, in learning about it, in actively participating in its restoration and management is also unclear, and in fact little data typically exists in the city to provide a comprehensive view of this biophilic engagement. These are important dimensions of biophilic cities, not least for the ways in which these biophilic activities and engagements help to build social capital and help to foster connections between and among people, and a larger sense of community.

Future Directions in Strengthening Oslo's Biophilic Commitments

Oslo has many impressive biophilic qualities, and it is hard to identify another city with as many natural qualities or assets. There is much to build upon here: very large blocks of natural forests seemingly nearby to every resident and neighbourhood, a beautiful fjord and aquatic areas nearby for fishing, swimming and boating; and a city with extensive parks and green spaces in between. It is truly, as the motto suggests, a green-blue city. And plans and planning standards are, for the most part, in place to ensure that these precious green qualities will endure: an impressive commitment to steering new growth into existing urbanized areas, and an almost sacred commitment to protecting the city's forested areas.

The city has, moreover, established impressive biophilic ambitions for the future, notably the daylighting and restoration of the main rivers and streams running through the heart of the city, and connecting in the future the forests and the fjords. This is visionary, bold, transformative, and has the potential to do more than any other strategy to enhance quality of life and help to (further) grow and nurture a biophilic population. It will also represent an exemplar for other cities to follow.

That said, there are many ways in which this green city could further strengthen its biophilic qualities. Partly, the challenge will be to enhance and restore the natural qualities in the densest and oldest parts of the city, and this will require a variety of creative biophilic design techniques and ideas not yet in wide use (for example, vertical gardens and green walls).

Further greening the more urbanized areas of the city, while at the same time pursuing further densification of the city, will be a challenge, but it also represents an opportunity for Oslo to further solidify its status as an exemplary biophilic city. Urban trees, rooftop gardens, bird- and wildlife-friendly building designs are just a few of the many things the city could do in these more intensive urban zones. Re-imaging the urbanized city as a bountiful city, and looking for every opportunity to insert and integrate food production is another opportunity: planting fruit trees in public spaces, encouraging balcony gardens and converting small urban spaces, where possible, to community gardens (for example, Nordalh 2009).

Partly, the challenge will be one of further fostering an outdoor culture, and finding new ways to nudge residents outside. It may be possible, moreover, to further deepen the sensory awareness and connections in this city: more work on the soundscape, more work in understanding the deeper ways in which residents interact with the outside environments; the breezes felt, the experience and enjoyment of winter weather and greater connections with the dramatic geology that forms this special place.

The new planning and design connections being forged with the coastline and fjord are impressive as well, evidenced in the innovative design of the Oslo Opera House and the coastal promenade and trail system envisioned to bring residents closer to the water-edge. Further efforts at enhancing visceral and emotional connections to the fjords and aquatic environments could be made, and is major challenge for the future. And Norway as a nation has a long seafaring tradition suggesting a special opportunity (and duty) to take steps to guard over and protect larger marine and ocean environments under its stewardship. Part of the task will be to educate about the biodiversity and immensely fascinating marine life nearby the city, and to find creative ways to make this life and biology more clearly visible to urbanites.

In other coastal cities, institutions such as urban aquaria have played an important role in fostering new ocean sensibilities and connections. A promising model can be seen in the work of the Oceanario, in Lisbon, Portugal, whose CEO, João Falcato, I recently interviewed. Oceanario is one of some 140 public aquaria in Europe, all with an educational mission. In the case of the Oceanario, its message of 'One Ocean' reaches a million visitors each year. While this facility looks much like other urban aquaria, much of its educational effort has focused on needed shifts in urban lifestyle and consumption. Most of the exhibits here emphasize what a visitor can do to reduce impacts on the ocean. Special efforts have been made to work with schools throughout Portugal and to help teachers educate on subjects such as overfishing. The Oceanario has even created a mascot, Vasco, who has become recognizable by kids throughout the country, a kind of superhero for the oceans. The Oceanario contains Vasco's house, which makes it fun to imagine ways to reduce impacts on oceans and to live more sustainably; from rooftop wind turbines and the water conservation in the bathroom, to the displays about local food in the house's kitchen. And the kids seem enthusiastically to be embracing the possibilities of these lower impact lifestyles.

It is also not clear how and in what ways Oslo expresses its commitments to nature beyond its boundaries. A city's biophilic sensibilities are tested by nurturing nature at home, but also by taking steps to protect and repair nature in other parts of the world. In some sense, this is about taking responsibility for the impacts of a large footprint, but it may extend behind culpability to a more proactive duty to support global conservation. In several policy documents there is acknowledgment of the large ecological footprint that Oslo exerts (indeed every affluent northern hemisphere city) on people, landscapes and nature thousands of miles away. To what extent could municipal procurement policy be reformed to avoid the most damaging impacts (for example, the purchasing of tropical hardwoods, or products with palm oil)? And what proactive initiatives might be supported (financially, technically, politically) in biodiverse and threatened places around the world? These are perhaps the less obvious ways in which biophilic sensibilities can manifest.

Overall, Oslo is already a very biophilic city, there can be no doubt. It can be especially proud of its work to protect forests, and to provide an unparalleled degree of access and proximity to wild nature – especially forests, shorelines and water. It already lives up to its motto of the blue-green city, and has set some extremely ambitious goals for the future, especially daylighting and restoring the remaining seven rivers flowing (at the moment in pipes underground) through the city. Implementing this goal would solidify Oslo's claim as Europe's (and perhaps the world's) Biophilic City.

Bibliography

Beatley, T. 2011a. *Biophilic Cities: Integrating Nature into Urban Design and Planning*. Washington, DC: Island Press.

Beatley, T. 2011b. 'Blue Urbanism', in *Places*, online journal, found at: http://places.designobserver.com/feature/blue-urbanism-the-city-and-the-ocean/26328/.

Beatley, T. 2009. *Low-Carbon Copenhagen*, report prepared for the Timelines Engineering, found at: Centre for Development and Environment, University of Oslo 2002. *The Ecological Footprint of the City of Oslo – Results and Proposals for the Use of the Ecological Footprint in Local Environmental Policy*, Western Norway Research Institute.

City of Oslo, undated. *Fjord City Plan*. Department of Urban Development, Oslo Waterfront Planning Office, Agency for Planning.

Earle, S. 2009. *The Planet Is Blue: How Our Fate and the Ocean's Are One*. Washington, DC: National Geographic Society.

Gehl, J. 2010. *People and Cities*. Washington, DC: Island Press.

Institute of Transport Economics, University of Oslo 2009. 'The Challenge of Sustainable Mobility in Urban Planning and Development in Oslo Metropolitan Area', July.

Institute of Transport Economics 2009. 'The Challenge of Sustainable Mobility in Urban Planning and Development in Oslo Metropolitan Area', Norwegian Centre for Transportation Research, July.

Kellert, S., Heerwegen, J. and Mador, M. (eds) 2008. *Biophilic Design: The Theory, Science and Practice of Bringing Buildings to Life*. New York: Wiley & Sons.

Oslo City, undated. 'Oslo Application for the European Green Capital', found at: http://www.oslo.kommune.no/getfile.php/byr%C3%A5dsavdeling%20for%20milj%C3%B8%20og%20samferdsel%20%28MOS%29/Internett%20%28MOS%29/Dokumenter/presentasjoner/Oslo_ECG_aplication.pdf.

Oslo City Council 2007. *Oslo Towards 2025: The 2008 Municipal Master Plan*. Oslo, Norway.

Oslo Green Capital 2011. 'Oslo Green Capital', slide presentation, undated, found: http://www.oslo.kommune.no/getfile.php/byr%C3%A5dsavdeling%20for%20milj%C3%B8%20og%20samferdsel%20%28MOS%29/Internett%20%28MOS%29/Dokumenter/presentasjoner/EGC_oslo12-01-09text.pdf [accessed: 1 May 2011].

Oslo Kommune 2008. 'European Green Capital Application from Oslo, Including Additional Information', 12 May 2008.

Oslo Kommune 2007. *Grontokan for Oslo*. Plan-og bygningsetaten, Oslo Kommune.

Oslo Kommune 2008. 'European Green Capital Application from Oslo, Including Additional Information', 12 May 2008, Oslo, Norway.

Nordahl, D. 2009. *Public Produce*, Washington, DC: Island Press.

Norwegian Institute of Transport Economics 2010. 'Health Effects of Increased Cycling for Transport in Selected Small Cities in Norway', Norwegian Centre for Transportation Research, October.

Sverdrup-Thygeson, A., Borg, P. and Bergsaker, E. 2008. 'A Comparison of Biodiversity Values in Boreal Forest Regeneration Areas Before and After Forest Certification', *Scandinavian Journal of Forest Research*, 23, 236–43.

Norwegian Institute of Transport Economics 2010. 'Health Effects of Increased Cycling for Transport in Selected Small Cities in Norway', *Norwegian Centre for Transportation Research*, October.

Wilson, E.O. 1984. *Biophilia*. Cambridge, MA: Harvard University Press.

Wilson, E.O. 1993. 'Biophilia and the Conservation Ethic', in *Biophilia: The Human Bond with Other Species*, edited by S. Kellert and E.O. Wilson. Cambridge, MA: Harvard University Press.

Wilson, E.O. 2007. *The Creation: An Appeal to Save Life on Earth*. New York: Norton & Company.

Chapter 2
Oslo: A City Framed by Forest

Karen V. Lykke Syse

Introduction

When I look out of my office window from the University of Oslo, I have a northerly view of an expansive green wood. In April it is speckled with a haze of purple birch branches which explode into lime-green foliage by the time May arrives, and even the sombre spruce will spurt marker-pen green shoots before June. During the summer months, the broadleaves take on a hue of dark green, awaiting September's first yellow birch leaves swirling about the air like coins. The orange foliage of October soon blows away and the cold and naked month of November is left to be framed by the ever present spruce on the horizon. The streets are grey, the days are short, and it is often dark when I enter the office and dark before I go home again. Then comes the frost; it arrives on a clear day and the forest changes: a scattering of white, bearing promise of more in weeks to come. An early sunset shines pink light and long shadows through the trees. Then one morning the ground is covered by a full blanket of snow. It may well thaw quickly in downtown Oslo, but once it covers the woods, it usually remains. In the winter months of December, January, February and March, the trees I see from my window have a Christmas-card-like icing, and the view sometimes lures me out of the office – out of the dry and static office waft, out of the wet and saline grey slush filling the streets of Oslo – and into the crisp cold air to find a white carpet of skiable tracks that run for miles and miles through 19 municipalities and five counties.

This 170,000-hectare large forest is Oslomarka, which surrounds the city of Oslo to the west, the north and the east, while to the south the city meets the sea. Fresh water lakes within the forest provide the citizens of Oslo with water, and rivers and streams running out of this forest and into the sea have given people power since medieval times. These same rivers were instrumental to Oslo's industrial past, charging the factory mills with energy, and floating timber from the forest into the sea using the rivers as highways. Although Oslomarka can be visually defined from my view (which is where the houses stop and everything seems dark, green and coniferous), lighter green tentacles of these woods, containing more broadleaves, seem to erase the beginning of the forest or the end of the town. It might look like a fuzzy border; yet recent legislation has definitely mapped the beginning of this sylvan Eldorado and the end of a sprawling suburban area. This mapping, along with protective legislation, will ensure that future generations will also be able to feel that they live in a green city with vast recreational woodland within

Figure 2.1 Svarttjernshøgda

easy access, rather than in a sprawling if somewhat forested suburb. Creating this border between the forest and the city was politically initiated as early as in 1938, yet it took 70 years for the Norwegian government to pass the legislation needed to finally enact Markaloven on 1 September 2009 (Bugge and Reusch 2010).

In the following chapter, the forest surrounding the city of Oslo will be presented in an historical context. This forest is a palimpsest, holding physical and cultural traces of the past, and although but a few of them can be explored here, the main narratives explaining its history will introduce the reader to what this landscape represents. It is a landscape shaped through human use throughout the centuries, yet the form of human interaction with the environment changed significantly, and particularly during the 20th century. Oslomarka is an area containing places people anchor both meaning and identities to, it is an area of importance to ideas of cultural heritage as well as to ideas of nationhood, and within the last few decades, it has also become an area important to biodiversity. Not only is Oslomarka as a place contested, but where it starts and where it ends is also in dispute. Before we start delving into these questions, however, we must define the term 'Oslomarka'.

A Norwegian Wood

Oslomarka is a fairly new term which evolved in the 1930s to encompass all the forested areas surrounding Oslo. Trying to write a brief history of such a vast area is close to impossible; it covers as much as 1,600 square kilometres. In the following, I will use three terms: Oslomarka, as defined above, and Nordmarka, which is arguably the first area which was used for recreational purposes by the population of Oslo. Nordmarka is the part of this forest that lies immediately north of the old city centre (see Figure 2.2). Finally, the colloquial expression which is used to cover this vast sylvan frame, encompassing both the terms Oslomarka and Nordmarka, is simply *Marka*, which will be used throughout unless a further specification is called for.

Marka has been owned by various people through the centuries, and giving a comprehensive history about the various owners is beyond the scope of this

Figure 2.2 Map of Oslomarka

chapter. In general terms, much of the *Marka* forest was owned by the Roman Catholic Church until after the reformation in 1536, when the land belonging to the Church was transferred to the king. Some of this land was let, most of it was sold off, and some of it was kept as state property. Some parts of *Marka* have been commons, and some have been in private ownership for centuries. Some areas have been purchased by the city of Oslo to ensure clean drinking water, recreational areas – and in the latest few decades – to preserve biodiversity. Both small farmers and large estate owners have had, and still have, their shares of this vast forest. To the west, there were (and are) many small landowners, while the northerly parts of the forest have mainly been privately owned as a large family estate, since at least the 1600s. The history of the forest lying further east and south is complex, as it too has been owned by the church, the municipality, the state, by several small landowners and a couple of big landowners. As these areas have been managed and owned by so many different people or institutions, the history during the last couple of centuries is perhaps even more chequered and difficult to convey, and perhaps not so interesting for the purpose of this chapter. Again, generalizing history is a hazardous exercise. Nevertheless, the largest landowner in *Marka* is the Løvenskiold family, who own about 110,000 acres or 43,000 hectares of forest in Nordmarka. This family has, through the patrimonial and matrimonial line, been in ownership of what was known as Nordmarksgodset (or the northerly estate) for about four centuries (Bull 1918, Syse 2000, Ulstrup 2006, Moland 2006, Larsen 1952).

Woodland in Norway is generally but not exclusively referred to as *utmark* – the opposite of arable land and lowland grazing which is called *innmark*. The comparable terms infield and outfield can be found in Northern England and Scotland, although outfields are seldom wooded there. Outfields and woods have been used for grazing and harvesting fodder to bring back to livestock housed at the home farm. Hunting, fishing and harvesting berries have undoubtedly also been an important resource for people living in the vicinity of the woods, and as such, the term *utmark* does not compare to the American term wilderness because it is a part of nature which is harvested and towards which people have a utilitarian perception. The saying '*Utmarka er innmarkas mor*'[1] explains the intricate connection and usage of the outfields in traditional Norwegian agriculture. Since animals grazed in the outfields, and since fodder like foliage, grass and heather was transported from the outfields back to the farm, a steady stream of energy was brought from outfields to infields. When the animals, during the course of winter, were fed by fodder harvested from the outfields, this energy was transferred to the farm in two ways: firstly, the animals would yield meat and milk, and secondly, they would provide manure, which in turn would enrich the grain harvest on the infields. Because of this, grazing or harvesting rights were very important to the farmers and smallholders who lived in the vicinity of the forest and these rights were not always tied to commons or ownership. On the other hand, there is an arguable difference between forest and woodland. There is little energy or grass to be harvested in an old coniferous forest. But if it is left unplanted, a freshly harvested part of the coniferous forest will become broadleaved woodland suitable for grazing before long. This elasticity in perception was more prominent in the days before plantation forestry, as it would take longer for a new generation of conifers to establish in pre-industrial forestry. Another point is that it is difficult to re-establish a forest in areas where animals, and particularly goats, are grazing. Because of this, animals have to be restricted if one wants to maintain a productive forest. Today this is not a problem, as there are hardly any grazing animals left in Oslomarka. Nevertheless, it is interesting to note that the names Nordmarka,

1 The outfield is the mother of the infield.

Oslomarka or simply *Marka* all reflect that this area historically was considered and named outfield – albeit wooded – rather than forest. There are areas and parts of *Marka* which have the Norwegian word for forest, *skog*, as for instance Kroksskogen, Hadelandskogen or Harestuskogen, but they are contained areas within the larger woodlands (*Marka*) (Olsen 1919).

Trading Timber

The Norwegian timber trade picked up in the 1600s for two main reasons. Firstly, hydro-powered sawmills made it possible to efficiently saw logs into boards, something which gave a better usage of each log, leading to higher profit. Secondly, there was a need for large quantities of wood to fuel ironworks, another new industry which became important in the 1600s and 1700s. Increased export turned forests into crucial economic resources.

In the 1700s, after the Great Northern War, Christiania (as Oslo was called between 1624 and 1925) was teeming with shipbuilding and trade. The economy boomed and the city was transformed into a central trading port. Timber had been an important raw material nationally, but to develop a growing international timber trade required large-scale investments. Organizing a workforce with loggers to fell the trees, timber drovers to drag the logs by horse to a river and floaters to float

Figure 2.3 Horses and foresters ca 1910

Source: Photo supplied by permission of the Oslo Museum.

logs to a port was only part of the venture. In addition, one needed engineers who diverted rivers and streams into timber-floating highways, and finally one needed a steady flow of buyers. This was an enterprise for merchants with logistic skills and capital. Whoever had the financial means to carry out this endeavour would be generously rewarded in the early modern period. The timber trade was so economically rewarding that a new social group grew up in Norway, coined '*planke-adelen*'.[2] On the other side of this new social scale, were the peasants and rural labourers who provided the needed manpower; becoming a forest labourer provided employment for able and strong young men throughout Norway. Along the Norwegian coast, men seeking employment outside the farm were recruited by the early fishing merchants, while the surplus of labour inland in Norway found their ways into the forests (Lunden and Brenna 2002).

As a broad generalization, the forest labourers in the 1800s could be divided into loggers, drivers and floaters. The loggers first chopped the trees down using an axe, then the branches and bark would be stripped off in the forest. Finally, they divided the trunk into logs. The logs would be hauled into a stack, left to be then taken over by timber drivers who pulled them by horsepower and placed them in piles along the roadside. Later, these fairly small piles of timber were then driven to the closest river, and finally the timber floaters would float the timber out of the woods and down to the closest sea port. Floating timber was extremely dangerous work; even though lakes and rivers were engineered to make the logs' journey as smooth as possible, they often became tangled, requiring the floaters to make sometimes-lethal crossings out onto the floating logs to free them.

Forestry work was seasonal, with transport and water in its multitude of forms as the key: the logs were most often harvested in early winter before the snow covered the lower trunks of the trees. The timber was then pulled out and onto the roadside. Once the ice covered the rivers, the logs were hauled onto the ice, awaiting the spring thaw. When the ice melted into flowing water, the logs were deposited with a sudden great splash and the floaters could get to work. Floaters and loggers were often seasonal labourers, and floaters in particular would often come from afar. Nevertheless, in areas owned by large estates such as Nordmarka, the owners organized a network of tenant farmers living within the forest since this provided the landowner with a steady supply of timber drivers. In order to drive timber, one needed horses, and in order to have horses one needed a farm where one could grow oats and other fodder. This in part explains the farms and smallholdings dispersed around *Marka*, as the tenants were by contract obliged to work for the estate owner. Smallholders and farmers who owned their own land would also sell their labour to larger estates during the winter months when there was little farm work. The forest was a lively peasant community in the outskirts of Oslo, and the sounds of bleating sheep, lowing cattle, axe blows, falling timber, sleigh bells and neighing must have made the forest seem both a populated and

2 The timber aristocracy.

Figure 2.4 Floating timber 1916

Source: Permission of Oslo City Archives.

animated dwelling place (Halberg 1993, Tveite 1964). However, it was not before the late 1700s that this rural place became an interesting place for town folk to be.

Urban Pioneers in the Forest

In the late 1700s, the perception of where Christiania ended and the countryside began was clearly defined. The town itself was a small urban cluster of buildings surrounding *Akershus* fortress, and outside town was the rural municipality of Aker. The Norwegian upper classes (inspired by Rousseau) had, as early as the 1750s, started to take an interest in these rural areas. Wealthy merchants would keep a house in the city and a farm in the country, and would go and live in the country during the summer months. The citizens of Christiania had grazing and harvesting rights, but no ownership in the areas to the north of Oslo. These rights – endowed to these citizens of Oslo by the crown – were necessary to maintain a supply of fodder for horses and to provide a certain degree of self-sufficiency (Moland

2006: 25). Nevertheless, there is also evidence of a certain longing and admiration for a rural way of life – the pastoral – in this period. Traces of these longings are evident through a small society of Norwegian intellectuals, called Det Norske Selskab,[3] which was inspired by the French Revolution. Det Norske Selskab had a firm belief that the true identity of the Norwegian national soul was contained within the free Norwegian farmer. The talks and meetings brought forward action, as a group of powerful if not exactly wealthy farmers and politicians formulated the democratic constitution of Norway, and Norway officially released itself from Danish rule in 1814 (Bull 1918).

A few decades later, artists and writers inspired by the Romantic Movement discovered the forested areas outside Christiania, perceiving them as wild and wonderful in the same way as Wordsworth and Coleridge were inspired by the English Lake District and the Scottish Highlands. Accounts from the mid-1800s show us how artists and writers head towards *Marka* in order to explore what was then conceived as a fairly exotic place. They would walk into the woods for the day or for even longer excursions, and write about their explorations. Writers who have given us an idea about everyday life in Nordmarka forest in the 1840s were Peter Christian Asbjørnsen (1812–1885), Jørgen Moe (1813–1882) and Bernhard Herre (1812–1849).

They were a group of friends who all cherished hunting and recreation in the forested areas around Christiania. They can be called urban pioneers of the forest landscape (Moland 2006: 48). Asbjørnsen was a Norwegian folklorist, well known among present day Norwegians as he recorded fairytales and stories about trolls and other supernatural beings in Norwegian lore. However, for the purpose of this chapter, it is worth noting two of his publications about *Marka*; one is called ‘En natt I Nordmarken’[4] from *Norske Folkeeventyr* (1843), published with Jørgen Moe, and another less well-known publication was published a decade later: *Om Skovene og et Ordnet Skovbrug*[5] (1855), which is an account of forestry as a scientific endeavour. Here, Asbjørnsen shows his concern for the, at times, ruthless deforestation taking place in *Marka* and other places, and calls for a forestry practice which takes regeneration of trees into account.

In ‘En natt i Nordmarken’, Asbjørnsen tells the story about one of his walks in the forest. The reader is taken on an excursion to places which are familiar to many present day Oslo citizens, and is introduced to the people who lived in this forest. Asbjørnsen goes fishing with Elias Fisher, he meets up with Pål Sandungen;[6] however the everyday lives of these people is only used as a framing device to narrate the stories these people told him about all the supernatural beings

3 The Norwegian Society.

4 A night in the northern forest.

5 About the forests and an orderly Forestry.

6 Sandungen is the name of a well-known farm situated centrally in Nordmarka. It is still possible to stop here for meals and refreshments.

Figure 2.5 Asbjørnsen in 1870

Source: Photograph by permission of the Oslo Museum.

of the forest: the nisse,[7] the hulder[8] and the trolls.[9] Asbjørnsen's stories about his travels and his collections of fairytales inspired other artists' and writers' romantic accounts of Norway in this period, and became part of what has later been coined 'the Norwegian National Project'.

Some of these writers were urban hunters, and they opened the door to everyday life in the forest to us through their memoirs. For instance, Fredrik Otto Juell (1803–1888) described how great the hunting was for grouse, black game and hare (Juell 1875). He was called 'Lynx-Juell' as he once shot a large lynx which was preserved and later displayed at the Natural History Museum in Oslo. There is still lynx in *Marka*; however, the bear and wolf which were still

7 A mischievous gnome-like creature.

8 A wood-nymph who lured men into the underworld.

9 Evil but fairly stupid ogres.

Figure 2.6 Angling by Sognsvann 1907

Source: Photograph by permission of the Oslo Museum.

roaming the forest in the early to mid-1800s became extinct by the 1880s. Another important hunter-come-writer was Bernhard Herre, who wrote about both the people who lived in the forest and about the forest itself. He was a keen hunter and trapper, and spent much of his time roaming around the forest pursuing game. He describes panoramic views and scenery, which are still easy to recognize when walking in *Marka*, and he expresses the feeling of solitude in the forest in a way modern readers can relate to. Herre shows both tenderness and affection in his accounts of the forest and the people within it. Having said that, Bernhard Herre depicts his journeys to *Marka* with less romance and more realism than what was typical for this period. Although his landscape descriptions might belong to the romantic genre, his descriptions of the people themselves belong to a social realistic tradition. On his trips he met the woodlanders: foresters, hunters, trappers, fishermen, shepherds and barefoot children. Bernard Herre does not attempt to hide the hard graft, poverty and distress in these people's lives; life in the forest was not for the lily-livered[10] (Herre 1850). It was mainly these common people writers like Asbjørnsen that Juell and Herre met and used as a starting point for narratives. However by the 1870s, a new group of people entered *Marka* (Bull 1918).

A Walk in the Woods

Like in some other European countries, the trend of going for walks in the countryside for pleasure became a trend in the late 1800s. The people (mainly belonging to the middle or upper classes) who partook in this new sport were called 'tourists'; in fact, going for a walk for no reason except to experience nature was called going for a *tur*[11] in Norwegian. The tourists came for the view and well-being, for fresh fragrant forest air, and for bodily exercise. Going for a walk on Sundays became an institution among the population of Oslo. As early as 1870, the Norwegian trekking association (DNT)[12] had marked a route through the forest. During the next 30 years, the network of marked tracks and trails through the forest increased, and it became more and more popular to go for forest walks. Through the popularity of the '*tur*' and through all the voluntary work from DNT, a whole new lifestyle developed; going for a Sunday walk in *Marka* became almost institutionalized among the middle and upper classes, and was perhaps a way of delineating the distinctions between the educated middle classes and the rising working class (Nedrelid 1993).

At the same time, another form of recreational pursuit took shape: skiing, which soon became the chosen winter sport of the growing urban population. Young men took to skiing first; after work or studies on a Saturday afternoon, they would head towards the forest. They might seek accommodation among tenant farmers who took in lodgers, or they would join the skiing clubs who erected cabins here and there. Skiing was by no means a new to Norwegians, but skis were perhaps considered more a means of transport, rather than a sport as such. The first skiing competition took place in Nordmarka in 1886 (Moland 2006: 75). Not long after, the national hero Frithjof Nansen crossed Greenland from east to west. After Nansen's return

10 Perhaps their misery somehow reminded him of his own misery – he came to a desperate end as he was found in the forest, killed by his own gun by a bullet to the head in 1849, only 37 years old.

11 From the French *Tour.*

12 The literal translation is The Norwegian Tourist Association, however, because of the modern connotations to the word tourist, the organization prefers to emphasize trekking rather than tourism.

Figure 2.7 Skiers 1890

Source: Photograph by permission of the Oslo Museum.

and publication of *På ski over Grønland* (On skis across Greenland), recreational skiing really caught on. Between 1886 and 1900, no less than 22 skiing clubs were established in *Marka*, and the Norwegian Trekking Association had started marking their paths with blue painted stripes on the trees for summer and red for winter.

1898 marked another important event for the population of Oslo, as Holmenkollbanen opened. This was a tram service running from Majorstuen, by the western centre of Oslo, and all the way up to Holmenkollen. At first, the final stop was at 242 metres above the sea level, and after an expansion in 1916 the tram-line could take the public as far as 277 metres above sea level. Holmenkollen is a very good starting point for walks and skiing trips alike, and gave the growing population of Oslo easy access to the forest. This fact was something used actively to market the tram way, in fact a whole new suburban area developed alongside the tram-lines (Nilsen and Nyquist 1972, Andersen 1993).

Marka had indeed become accessible to all, and the next few decades led to even more signposts, improved paths and skiing tracks within the woods. For the population of Oslo in the late 1800s and early 1900s, *Marka* was a place they could explore both as a traditional rural area and as a place they perceived as wild and

Figure 2.8 Horses and skiers by the tram-lines, 1927

Source: Photograph by permission of the Oslo City Archives.

untarnished by urban industry. They would meet grazing cattle and sheep during the summer months; they might bump into smallholders, shepherds, foresters and hunters. When the snow covered the ground and they went skiing, they might do this to the sound of sleigh bells and snorting horses. The sounds and smells of a living woodland community only a tram ride away from industrial, overcrowded Oslo was a tangible reason to pay the forest a recreational visit.

During the first three decades of the 1900s, life for the woodlanders was in many ways a continuation of the way life had been led for centuries. At the same time, their vicinity to Oslo meant they were influenced by the needs of the urban population in more ways than opening their homes for accommodation and meals during the weekends. As the growing population of Oslo needed more water, some huge engineering projects changing the course of waterways were initiated to supply water to the city. Securing the drinking water supply of Oslo also led to an earlier termination of timber floating in Nordmarka, compared to the rest of Norway. Safeguarding the drinking water for the population led to road constructions to compensate for the rivers as alignment routes, an engineering exercise which, paradoxically, culminated in the first major conflict of interest between foresters and the recreating public, a matter I shall deal with below.

Mechanization, Technological Invasion … and Even More Recreation

The first chainsaws which could be operated singlehandedly came into general use after the Second World War. Alongside them came motor vehicles. It became necessary to build forestry roads to accommodate timber lorries which would take the timber out of the forest and down to the mills. A new network of timber roads slowly but steadily intersected the forest and the timber roads were both a blessing and a nuisance: the forest became more accessible, and fewer areas of the forest could be regarded as 'inaccessible wilderness'. At the same time, the forest became depopulated. In 1900, there were 600 workers labouring in the forest. In 1950, this number had decreased to 200, and in 1999 there were merely four labourers left employed in *Marka* (Lange 1966: 41, Syse 2001). Although the roads were strictly for timber lorries, and driving motor vehicles was reserved for foresters and the local population who lived in the forest, the timber roads also became an entranceway to a new group: cyclists.

Figure 2.9 Skiing elegantly in 1905

Source: Photograph by permission of the Oslo Museum.

From about 1900 to 1930, the popularity of walking and skiing in the woods increased at a staggering rate. Class and gender limitations were dissolved, and both women and working classes poured into the forest. Scouts, guides, students, various sports clubs, gentlemen's clubs, labour associations – *Marka* was large enough to absorb and welcome everyone. Walking and skiing in *Marka* became the Oslo population's favourite healthy pastime, helped on by a variety of organizations through even more signposted routes, through publicity campaigns and through socialization. The Labour Party were keen to promote the health benefits of walking and experiencing nature, and the politician Einar Gerhardsen's (1897–1987) memoirs describe many trips to *Marka* with friends, family and fellow politicians (Hodne 1994, Moland 2006).

Cabins were built to accommodate both day visitors and overnight guests, and some people wished to have a cabin of their own in the forest to ease the logistics of recreation. In 1934, an important event marked the importance this new recreational pursuit had for the population of Oslo. A property within the forest was being sold off to the tenant farmer, and there was a public outcry due to concern about what would happen to this property. Would this farm in due course be chipped into lots and sold to people wanting to build their own cabins? The potential problems relating to property, ownership and private enterprise led to the founding of *Oslomarkas Friluftsråd*,[13] an organization who would speak with the unison voice of the people who loved the woods, irrespective of who owned it (Berg 1952).

In the 1930s, the science of forestry had come into existence. In previous centuries and decades, forestry had been a matter of harvesting what was already there, and leaving the forest to naturally regenerate for future generations. Now, new ideas of rationality entered the woods, and two main features were typical for this period. One was plantation forestry; the other was clear-felling trees. By clear-felling trees, the felling process itself became both more efficient, and the regeneration or rotation of trees would take a shorter time. Clear-fells simulated what happens after a forest fire, however, skipping the first ecological processes where pioneer species encroach on the area before finally the shade-loving Norway spruce regenerates and regains ascendancy. Rather than waiting for the first part of this ecological process to happen, the clear-fell would be planted with neat rows of seedlings. By planting the trees a bit too close for comfort, one ensured straight growing trees with few lateral branches, thus optimizing the quality of the log (Syse 2000: 53). Once the trees had started growing, their dense growing conditions required the trees to be thinned. Since the trees grew in straight lines, thinning could proceed mechanically.

Mechanization required tractors which in turn required roads – not just roads for timber lorries, but roads for forest tractors. The network of roads increased at a staggering rate. In 1930, the total length of roads in Nordmarka was 95 kilometres and by 1962, 300 kilometres of roads had been built (Lange 1966: 104). The forestry roads were not uncontested; however, the first and most important conflict regarding large engineering projects running through *Marka* came in 1946. The owner of Nordmarka estate, Carl Otto Løvenskiold, wished to protect his forest against the onslaughts of high power masts and large roads, which had been started by the Nazis who were governing during the war years, and which were planned to be continued

13 Oslo Regional Council for Outdoor Activities and Recreation.

by the post-war politicians. In 1946, the Oslo City Council voted for a massive row of electricity masts running straight through central Nordmarka. The press were surprisingly uninterested in this, and it was Løvenskiold who voiced a clear protest against the municipality's plans. A huge PR-campaign was organized and in November 1946 no less than 30,000 people marched the streets of Oslo to protest. To this day, it was the largest organized political demonstration in post-war Norway. The years following this demonstration and leading up to the 1970s were characterized by a series of technocratic suggestions stemming from various officials in the city of Oslo: the municipal forester wanted modern forestry; Oslo waterworks wanted to modernize the water supply; Oslo highway department wanted to build new roads (for general use); and the railway wanted to run tracks through the forest (Moland 2006: 139). All of these technical innovations would somehow alter the forest, and the recreational groups were a strong voice against the proposals. Meanwhile, city planners in Oslo developed the 1950 General Plan of Oslo, which will be discussed by Kine Halvorsen Thorén later in this volume. The need to retain a green structure both within the city itself and to somehow protect the surrounding green belt was, if not agreed, than at least placed on the agenda. The remaining question was really how to implement the General Plan; which areas should be protected against the needs to accommodate an expanding city?

Forestry was still a profitable industry, and the trees had a tangible commercial value. Nevertheless, in the mid-1970s, all timber floating in *Nordmarka* came to end, and by 1980, even local transport-floating of timber from one side of a lake to another stopped. Still more roads were needed to replace the role of the river systems, and this was criticized by recreational groups. The voice of recreational groups became strident in the debate about *Marka*. In 1972, another large demonstration took place, this time in the forest itself. The local paper Nordmarksposten's ironic comment was that only two of the participants had come to the demonstration venue without using the forestry road to get there. Several polls carried out in the 1970s proved that *Marka* was definitely important to the growing population of Oslo (see Lind 1973a, Haakenstad 1972, Haakenstad 1975). The reports concluded that high technology and modern clear-cut forestry upset the recreational interests. Furthermore, the large clear-cut areas were criticized by biologists and nature conservancy agencies that were concerned by the lack of biodiversity the monoculture clearly produced. In particular, lichens and moss seemed to be disappearing from the woods. This debate in many ways shaped the condition for present-day forestry in *Marka*. Biodiversity set aside, the recreational stakeholders had an ambivalent view of these clear-fells that were shaped by seasonality (Lind 1973b: 52). Although the felling sites were impossible to combat in the summer months due to branches and debris from the trees, in winter-time, once the snow covered the branches and tree stumps, the felling sites were sunny and open spots that were great to cross on skis as they opened up the view. A similar ambivalence can be traced with relation to the forestry roads, as touched on above. Although people disliked meeting timber lorries or the local forestry population's cars along the roads, these same roads made the forest accessible to groups who hadn't previously been able to use the forest. Cyclists, disabled people in wheelchairs and young people pushing prams and buggies found the forestry roads very convenient (Syse 2000: 63). The forest was truly there for all to enjoy.

Nature and Nationhood

What was it that made the population of Oslo such keen walkers and skiers? Did living in the vicinity of this green forest shape their behaviour, identity and ideology somehow? We have already touched upon the importance of nationhood and national heroes such as Frithjof Nansen. Nansen established the term *Friluftsliv*, or outdoor culture,[14] in the Norwegian ideology, and through his

Figure 2.10 Fritjof Nansen

14 A literal translation is free-air-life. The term covers any recreational pursuit carried out in natural surroundings, such as walking, mushrooming, hunting, (rough) camping, trekking, skiing and so on. Some of these activities, like skiing, are activities or sports in their own right, however they still fit under the conceptual umbrella free-air-life.

life as an explorer and national hero he became an important example for future generations of Norwegians.

He summarized his thoughts on being at one with nature in a speech titled Friluftsliv, held for the youth chapter of Den Norske Turistforening[15] (DNT) in 1922. The speech was reprinted in the encyclopaedic work, *Oslo Marka*[16] (Larsen 1952) and given a prominent position on page one. He writes:

> The Norwegian youth of the future ought ideally be a friend of nature, a sportsman, a free, independent man who knows what he wants, with a hardy spirit and a strong body – who knows himself, his people and his country, who asks much of himself and demands little of his surroundings. What replenishes us and takes us back to a more humane way of life is a simple life led in nature, in forests, mountains and moors, in a vast solitude, where new and larger streams of conscious thought comes to us and brands us in ways that can only be erased by effort. (Author's translation)

Frithjof Nansen ends his passage by calling for independence and self-reliance as virtues every Norwegian man and woman ought to strive for. The concluding message is that these virtues can be found and nurtured through a simple life led in nature. Frithjof Nansen had been a champion skier in his youth, and was a renowned Norwegian explorer, scientist, diplomat, humanitarian and Nobel laureate; he obviously had a strong impact.

During the occupation of Norway in the Second World War, young men took Nansen's words to heart. Young boys and men in the Oslo area fled to the woods, and led the local resistance fight between the trees in an area they knew a whole lot better than the invading German forces. They were coined 'Gutta på Skauen'[17] and fought for freedom after Nansen's calling. Provisions, guns, radios and equipment to sabotage German activity were dropped into the woods by British airplanes and picked up by the local resistance groups. During the war years, the forest as an arena for harvesting food was rediscovered. Picking berries and fishing in the lakes became a means to put food on the table for many people, and wild food such as sorrel and nettles were revived during the war years. The woods were also used as a safe passageway, if not totally away from the control and restrictions of the German soldiers, than at least with less of a chance of getting caught out or questioned about one's errands (Christensen 1993). As such, the forest became a free zone, and perhaps also a landscape associated with heroes. Heroes of another kind also emerged from these woods: the foresters, who had used skis as a means of transport all their lives seemed almost effortlessly to harvest gold medals in both national and international competitions such as the winter Olympics. An

15 The Norwegian Trekking Association.

16 The first, but unobtainable edition of this work was published in 1939 by the same editor as the edition above.

17 The forest lads.

example of this is Johan Grøttumsbråten, a forester from Nordmarka who won gold medals in the 1924, 1928 and 1932 Olympics, and who worked as a courier for the resistance movement during the war years. He was one of several skiing-heroes-come-resistance-heroes who emerged from these woods; friluftsliv in *Marka* was certainly rewarding both for body, soul and for feeling of nationhood.

Marka was a free zone, a place one could escape to. One could escape the city, one could (somewhat) escape modernity, and one could escape the Nazis. However, *Marka* was a free zone for other reasons too. In the article 'En morsom og uskyldig tid'[18] the ethnologist Liv Emma Thorsen describes how Oslo's teenagers used Nordmarka in the post-war years. *Marka* was a place were friends of both sexes met up and became visible to one another. The little lakes were great for both swimming on a hot summer's day and for fishing. Bonfires and tents provided

Figure 2.11 Swimming in the woods

Source: Photograph by permission of the Oslo Museum

a romantic setting, and the forest became an arena which satisfied the needs for teenagers to be able to retreat and interact without too much adult intervention (Thorsen 1994). Later research shows how the rural youth in adjoining areas also had a need to escape into the liberty of the forest (Syse 2000).

18 A fun and innocent time.

1980–2000 Backdrop Forestry

During the last couple of decades in the 1900s, major changes took place in Norwegian forestry. Animals were by and large excluded from the forest, and forestry became thoroughly industrialized and mechanized. Simultaneously, an awareness of the forest's intrinsic values came to the surface, and particularly the values of biodiversity. As many as 22,000 kinds of plants and animals are associated with Norwegian forests, and no less than 900 of these species are rare or endangered. This concern with biodiversity in scientific communities and environmental movements has forced both the government and the forestry owners to take action. Foresters had to maintain and encourage the biological values of the forest, while at the same time retaining the forest's recreational values. Both of these concerns led to a new legislation which was passed in 1993, and which limited the manner in which forestry practice could be carried out in the vicinity of urban areas, focusing on maintaining both the recreational and aesthetical values of the forest. The foresters had to think about aesthetics as well as biodiversity when they felled trees – a practice coined the 'backdrop forestry'. Further regulations were passed, demanding that the foresters limit the sizes of the clear-fells, that they used natural regeneration of trees where this was suitable, that they stopped planting trees in neat rows, that they kept the felling sites along the natural contour lines and that they stayed away from square sites. Another thing was what a forester tongue-in-cheek called 'hiding the whole thing from the tourists'. Alongside lakes and streams, the trees were left to maintain the views and provide game with natural shelter. Since most people now entered the forest from the roadside, the foresters 'hid' the felling sites by leaving the trees closest to the roads unfelled until the trees behind them had grown to a certain height and thus made the clear-fell less aesthetically challenging. These alterations of forestry practice guarded by a vigilant public seemed to appease most of the interest groups at the turn of the millennium.

Planning for a Green Oslo

As mentioned in the introduction to this chapter, the question of where the borders of *Marka* actually were was raised as early as in 1938. This was when Oslo og Omland Friluftsråd[19] (OOF) suggested a border between the city and *Marka*, and the discussion regarding where exactly this border ought to be, and what ought and ought not be allowed within this border has been a complex and ongoing discussion with many stakeholders taking part. Two of the five counties and 18 of the 19 municipalities have collaborated in this work through OOF. The main agenda for OOF has been to protect *Marka* and other green areas, and to ensure accessibility through so-called 'green corridors' which are green passageways from the urban and suburban areas and into the forest. As mentioned above,

19 Oslo Regional Council for Outdoor Activities and Recreation.

the architects behind the 1950 General Plan of Oslo sketched a map of Oslo, emphasizing the importance of green corridors running between the urban areas, through the suburban areas and into *Marka*. OOF also encourages people who are unfamiliar with outdoor recreation to take it up. One of the many arguments behind this is that outdoor recreation in nature is proven to be beneficial to physical health, mental health and general well-being (Pretty et al. 2005, Bedimo-Rung et al. 2005, Howard and Frances 2006). Because of this, they believe people ought to be encouraged to explore the forest in both summer and winter. Organizations such as OOF and DNT encourage the public to walk and exercise in the local area, and even the Norwegian government actively encourages outdoor recreation through various ministries.

After many years of discussion, the green paper suggesting a new set of forest regulations formalized in law as the Markalov and passed in 2008. The purpose of this new legislation is to ensure and advance the public's possibilities for outdoor recreation, nature experiences and sport. Markalov will safeguard the borders of *Marka*, and will ensure the conservation of a rich and varied landscape and the natural and cultural heritage herein. Although organized sports such as skiing, biking or trekking are to be considered and promoted, the intrinsic intangible values by merely experiencing nature are to be endorsed. Simultaneously, the legislation

Figure 2.12 Mushroom hunting, 1869

Source: Photograph by permission of the Oslo Museum.

must take into consideration the sustainable use of the forest for hunting, fishing and forestry. The object clause of Markaloven is to let everyone's voice be heard.

This brings us back to Bernhard Herre, as almost two centuries ago he conveyed the voice of people leading their everyday lives in the forest. He described hard graft, barefoot children picking berries, hunters and trappers harvesting game and smallholders and foresters squeezing their livings out of the woods. The situation is very different today. Rather than representing hard graft and multifarious ways of making a living through natural resources, *Marka* provides an invigorating experience; a place city dwellers go, where they can exercise their limp paper-pushing limbs in fresh air to keep somewhat fit. While writing the final lines of this chapter, I received an invitation to a University of Oslo social event: mushrooming in *Marka* with colleagues from all the different university departments, led by a biologist who is an expert on *fungi*. It is a sunny autumn's day and the view from my office window yet again pulls my attention away. The last weekend of the crayfish season is coming up and the birch's golden coins have set the forest on fire. The intoxicating smell of damp earth and rotting leaves is just an eight-minute metro ride away and I know of a lochan (a small lake) with ebony-coloured noble crayfish. I've signed up for the mushrooming event, but who knows, on my way into the depths of these woods to find that magic lochan I might come across a patch of *chanterelles*. I keep a basket in my office for immediate autumnal escapes such as this and my colleague in the office next door keeps a pair of cross-country skis behind his door. He will have to be a little patient yet, but *Marka* holds possibilities pivoting around the changing seasons. The keynote for all stakeholders' voices is conservation through use, consideration for different interests, and access for all.

Bibliography

Andersen, B. 1993. *Holmenkollbanen: Kort historikk fra 1898 til 1993*. Oslo: Lokaltrafikkhistorisk forening.

Asbjørnsen, P.C. 1855. *Om Skovene og om et ordnet Skovbrug i Norge*. Christiania: Steensballe.

Asbjørnsen, P.C. and Moe, J. 1843. *Norske Folkeeventyr*. Christiania: Dahl.

Bedimo-Rung, A.L., Mowen, A.J. and Cohen, D.A. 2005. The significance of parks to physical activity and public health: A conceptual model. *American Journal of Preventive Medicine*, 28, 159–68.

Berg, O.J. 1952. Tursporten i Oslomarka, in *Oslo-marka: Nordmarka, Østmarka, Lillomarka, Sørmarka, Vestmarka, Krokskogen, Romeriksåsene*, edited by H.T. Larsen. Oslo: Halvorsen & Larsen forl.

Bugge, J. and Reusch, C.H.P. 2010. *Håndbok for markaloven*. Oslo: Oslo og omland friluftsråd.

Bull, E. 1918. *Akers historie.* Christiania: Olaf Norli.

Christensen, T. 1993. *Marka og krigen: Oslomarka 1940–1945*. Oslo: T. Christensen.

Haakenstad, H. 1972. *Skogbehandling i et utfartsområde: En opinionsundersøkelse om Oslomarka*. Meldinger fra Norges landbrukshøgskole. Ås.

Haakenstad, H. 1975. *Skogskjøtsel i rekreasjonsområder: Skog og friluftsliv i to modellområder i Oslomarka*. Ås: Norges landbrukshøgskole, Institutt for skogskjøtsel.

Halberg, P.T. 1993. *Den stolte sliter: Skog- og landarbeiderne 1900 til 1990: En kamp for likeverd*. Oslo, Fellesforbundet, Seksjon skog og land.

Herre, B. 1850. *En Jægers Erindringer*. Christiania: Mallings Forlags-Boghandling.

Hodne, Ø. 1994. *Folk og fritid: En mellomkrigsstudie i norsk arbeiderbevegelse*. Oslo: Novus.

Howard, M. and Frances, M. 2006. Tasting the Waters: Discriminating Identities in the Waters of Blue Mud Bay. *Journal of Material Culture*, 11, 67–85.

Juell, F.O. 1875. *En gammel Jægers Meddelelser*. Christiania: Abelsted.

Lange, O.L. 1966. *Nordmarka og Nordmarksgodsets historie*. Oslo: Dreyer.

Larsen, H.T. 1952. *Oslomarka: Nordmarka, Østmarka, Lillomarka, Sørmarka, Vestmarka, Krokskogen, Romeriksåsene*. Oslo: Halvorsen & Larsen forl.

Lind, T. 1973a. *Publikums bruk av Oslomarka: Sammendrag av rapporter fra seks undersøkelser*. Oslo: Norsk Institutt for By- og Regionforskning.

Lind, T. 1973b. *Skogsdrift og friluftsliv i Oslomarka: Skogsdriftens konsekvenser for naturopplevelse og friluftsliv*. Oslo: Norsk Institutt for By- og Regionforskning.

Lunden, K. and Brenna, Å. 2002. *Frå svartedauden til 17. mai: 1350–1814*. Oslo: Samlaget.

Moland, T. 2006. *Historien om Nordmarka: Gjennom de siste 200 år*. Oslo: Christiania forl.

Nedrelid, T. 1993. *Ut på tur: På nordmanns vis*. Oslo: Cappelen.

Nilsen, K.A. and Nyquist, F.P. 1972. *Nordmarksboka*. Oslo: Grøndahl.

Olsen, M. 1919. *Om optegnelse av stedsnavne og andre sprogminder*. Christiania: Norli.

Pretty, J.N., Network, C.R. and Staff, C.R.N. 2005. *A Countryside for Health and Wellbeing: The Physical and Mental Health Benefits of Green Exercise*. Countryside Recreation Network (CRN).

Syse, K.V.L. 2000. *Lende og landskap: En analyse av skogens fysiske landskap og landskapspersepsjon i Nordmarka fra 1900 til 1999*. Oslo: Universitetet i Oslo.

Syse, K.V.L. 2001. Ethics in the Woods: Ethics, *Place and Environment* 4, 226–34.

Thorsen, L.E. 1994. *En morsom og uskyldig tid – skisse av en ungdomskultur*. Dugnad: Tidsskrift for etnologi.

Tveite, S. 1964. *Skogbrukshistorie*. Oslo: Skogforlaget.

Ulstrup, E. 2006. *Stuene i marka*. Oslo: Schibsted.

Chapter 3
Recollections of an Urban Green: The Frogner Park

Fredrik Chr. Brøgger

The subject of this chapter is the Frogner Park in Oslo, parts of which are also called the Vigeland Park, situated in the west-end borough of Frogner. It is a unique park in Norway and rather exceptional also in an international context, since its central part constitutes the vision and creation of a single artist, the sculptor Gustav Vigeland. The story of the area of Frogner stretches back from agrarian origins in the Middle Ages to its subsequent development as a bourgeois estate; in its modern park version it is intimately linked to the social history of Oslo and to 20th-century ideas of urban planning and recreation, landscaping and park construction and popular culture and the fine arts. My chapter deals primarily with the significance of the Frogner Park as an urban green, which is closely linked to the issue of garden design. My inroad into this discussion of the meaning of green urbanism is, however, a deeply personal and phenomenological one. In the first part of this chapter, I will share with you some of my own experiences, recollected in tranquillity, of what it felt like to have the Frogner Park – this large, green lung of the city of Oslo – as one's primary playground. I seek to evoke in concrete, empirical terms, how the park affected an impressionable boy between the ages of six and ten, and to sketch, in broad strokes, the sensations that the park provided me with. These constituted the experiential ground, as it were, from which my subsequent adult conceptions of, and attitudes to, nature developed. In the second, more theoretical, part of my chapter, I will discuss some of the biological, psychological and environmental implications of being surrounded with all this green stuff at one's most impressionable age.

My Childhood Recollections of the Frogner Park

In 1945, just a couple of weeks before Norway's liberation from the German occupation, I came into the world, which was that of the city of Oslo. This was not my fault, obviously; my patriarchal line was in fact the one to blame – Oslo people, the whole lot of them, ever since the 1700s. My parents' outlook was the same as that of most inhabitants of Norway's capital: one could of course travel to other parts of the country for vacation purposes and the like, but one would not want to settle down there. Thus I spent, more or less by accident of birth, my first

20 years in Oslo, more precisely in the lower end of Kirkeveien, right across from the large park at Frogner.

In my childhood, Frogner was not the 'yuppie' neighbourhood that it is today; it appeared a little worn and was peopled with punctilious, refined old ladies and gentlemanly, respectable burghers. My own family however – although of considerable bourgeois distinction in former days – was less than prosperous and existed, as it were, at the fringe of this respectability. My father, a literary critic and author, divorced my mother, an actress, when I was four, and I grew up with my father (not quite *comme il faut*, either, at this time) who remarried when I was seven. He kept living according to the dictates of an artist's lifestyle, which meant that his access to roast beef and apple pie, not to mention beer and wine, fluctuated with the fees from editorial offices and advances from publishers. In my mind's ear I can still hear, through the closed door of our dining room that was also my father's 'office', the keys of my father's manual typewriter hitting the sheet of paper like intermittent bursts of crackling hail.

For me, the sound of my father working did not, to put it mildly, captivate my interest. The Frogner Park, on the other hand, did. By 1947 it had, by and large, been completed in the form that it has today, and was awaiting my childhood daily visits in the first part of the 1950s. The American nature writer Wallace Stegner has said about the river valley of his Saskatchewan prairie childhood that it taught him 'about safety': as Stegner puts it, 'in a jumpy and insecure childhood [...] sanctuary matters' (*Wolf Willow*: 22). The Frogner Park offered me a kind of natural sanctuary, not so much from the urban hubbub of Oslo, which I have never found to be particularly threatening, but from, I suspect, a sense of being in the way at home, where my father worked at his typewriter at all hours, and where I had very little to occupy my restlessness. Thirty yards away however – on the other side of the street of Kirkeveien with its tram tracks in the middle – lay the park.

Frogner Park became my playground, my refuge, my enchanted forest. My childhood and the park are intertwined with unbreakable threads. I jumped on and off its merry-go-round and seesaws; I bounced rubber balls against the square bases of the Vigeland statues; I climbed the trees at Gratishaugen (the 'Free-of-Charge-Hill' from which track and field events could be watched); I threw snowballs in the winter at the naked statues (one point per breast, two points for the groin regardless of its sex, and three points – for whatever reason – for the head); I climbed up and balanced on the stone balustrade all around the Monolith – to mention just a few of my park activities. First and foremost, and far more importantly however, I ran around watching the life of all the animals that filled the various biotopes of the park.

When we are children, we are spontaneously aware that we share our lives with other living creatures. All over the park were different forms of life that crawled, buzzed, ran, flew and swam around me. Like most large, green spaces in urban areas, the park at Frogner was, and still is, a surprisingly biodiverse environment. Here were earthworms and spiders and millipedes, and flies and bumblebees and butterflies, and mice and rats and squirrels, and sparrows and wagtails and pigeons, and crows and magpies and gulls, and ducks and swans and fish, and dogs and cats and porcupines. To the small child that was me, they were not lower on the ladder than myself, they were simply other living beings sharing the space with me, ceaselessly fascinating – in the earth, in the grass, in the trees, in the air, in the water. The park teemed with beings small and large that appeared far more exciting, most of the time, than all the *homo sapiens* that hurried up and down Kirkeveien and, especially on Sundays, wandered about in the park.

Similarly, the art works in the park, for instance its more than two hundred statues, held little fascination for me, whether in summer or in winter – although the kind of Vigeland statues that I today appreciate the least, those massive stone creations of large muscular naked men and women, offered at least (unintentionally, one would think) some element of fun, because you could climb them. In the winter there was, however, something almost comic about these naked figures standing there with snow on their heads, on their shoulders, on their feet and on their penises (that did not even shrink in the cold). To me the numerous statues were simply there like everything else in the park, only more boring than the rest of it. They smelled nothing, were hard and stood stock-still. There was not much enthrallment to be had from Vigeland's stone-cold monuments, nor from the monumental axis-oriented design of the park itself.

If I acquired a sense of aesthetics at that age, it was more connected with the art of nature than the art of sculpture or garden architectonics. There were many things I found, little as I was, quite beautiful: the view of the buds of blue anemones forcing their way through the last wet and grainy snow in the grove of deciduous trees behind the Monolith; the sight and smell of the chestnut and lilac flowers in different places in the park; and, of course, all the summer shades of colour in the triangular and rectangular flower beds between the Vigeland Bridge and the fountain: red roses, white roses, red-and-white-striped roses, yellow roses, red-yellow roses, rosy roses, pink roses, orange roses, violet roses, plum-red roses, burgundy roses and black roses. To me the most awe-inspiring sight may nonetheless have been the flight of the ducks as they came sweeping in towards the lower Frogner pond in a wide curve, barely clearing the treetops that lined its shores. I never grew tired of watching them: the way they came sailing through the air while adjusting the curvature of their wings and slowing the speed of their descent until finally ploughing their way through the water and coming bobbingly to rest, quacking loudly to announce their arrival.

It was astoundingly beautiful, every time.

I did not know it at the time, of course, but my use of the park harmonized much better with the ideas of Marius Røhne, the first prominent landscape architect in Norway, than those of Gustav Vigeland, the nation's greatest sculptor. In the late 1910s and early 1920s, when the city of Oslo contemplated the future role of these areas for park purposes, Røhne and Vigeland clashed violently in the public debate over the use of the park. In connection with the city's discussion of this matter in 1919, Røhne 'made it clear to the Park Committee's President Mrs. Fernanda Nissen that he feared that Vigeland with his immense productivity might end up transforming the entire park into a sculpture park'[1] (Røhne 1965: 155). As Oslo's first publically appointed gardener, Røhne served as the head of the city's newly established *Parkvesen* [the Park Service, today entitled 'Oslo's Agency for Outdoor Recreation and Management'], a position he held from 1916 to 1948. He played a leading role in the movement of landscape architecture in Norway. In 1917, he authored a work document for the Park Service entitled *Den grønne by* ['The green city']. He regarded Oslo's public parks not primarily as places for the city's bourgeoisie to promenade in, but as 'folkeparker' ['people's parks'] – public recreation grounds to be enjoyed by all. As Røhne notes in his book *Oslo kommunale parker og grønnanlegg 1810–1948* ['Oslo municipal parks and green constructions 1810–1948'], to denominate the parks as the city's 'lungs' presupposes that as many people as possible make use of these 'helseheimer' ['homes of health']: 'One can hardly imagine a more quintessential communal possession than the parks. If one deviates from this conception and lets the parks serve special purposes at the expense of communal ones, the people's park ceases to exist'(Røhne 1967: 111).[2]

To Røhne, it was essential that parks be designed as natural environments: 'It is the identification with nature together with the natural endowments of the park that is the decisive factor' (Røhne 1967: 111).[3] In this respect, Røhne was clearly influenced by the 19th-century American landscape architect Frederick Law Olmsted, most famous for the design and construction of Central Park in New York City. Olmsted emphasized that parks should be designed for the public's use according the prevalent 'natural conditions, vegetation, water, and terrain of the place in question'[4] (Riksantikvaren 2009: 23), and Røhne similarly stressed that the Frogner Park ought to be designed as a nature commons. Vigeland, however, envisioned it mainly as a park for the celebration of his own art. According to a present-day article on the park's history, 'Today the whole world can enjoy the fact that Vigeland won the battle over the park'[5] (Haug 2010: 3), but my reasons for enjoying the park as a child were much more closely related to the objectives of Røhne than those of Vigeland. Røhne envisioned, in 1917, the Frogner Park as an environment to be designed according to its natural characteristics: 'One must realize that the terrain at Frogner must be considered nearly ideal; from nature's hand there is nothing stingy about

1 [Norwegian original:] 'Under behandlingen av saken gjør bygartneren [Røhne] parkutvalgets formann fru Fernanda Nissen oppmerksom på at han fryktet at Vigeland med sin umåtelige produksjonsevne ville komme til å omskape hele parken til skulpturpark.'

2 [Norwegian original:] 'En kan neppe tenke seg et mer utpreget felleseie enn parkene. Brytes det med dette og lar en parkene tjene spesielle formål på fellesskapets bekostning, opphører folkeparken … '

3 [Norwegian original:] 'Det er naturfølelsen sammen med de naturlige virkemidler parken har som er det avgjørende'.

4 [Norwegian original:] 'Utformingen av Olmsteds parker bygde på forutsetningene i naturforhold, vegetasjon, vann og terreng på stedet'.

5 [Norwegian original:] 'I dag kan en hel verden glede seg over at Vigeland vant kampen om parken'.

its variation, quite the opposite. The terrain is richly endowed with open flatlands, hillocks, and dells that rightly used represent the best foundation for a first-class park construction' (Røhne 1967: 151).[6] Indeed, when I roamed the park as a child, I was more preoccupied with the older sections ['Gamleparken'] that Røhne had helped design in the 1910s, than with the sections later developed according to Vigeland's plans. As Yi-Fu Tuan (1974) notes generally about Western garden designs – a description that perfectly fits Vigeland's baroque vision – they cater:

> primarily to sight. Of human senses sight is the most discerning spatially: the habitual use of the eyes leads us to appreciate the world as a spatial entity of well-defined lines, surfaces, and solids. The other senses teach us to perceive the world as a rich unfocused ambiance. Neither the formal gardens of the seventeenth century nor the 'natural' landscapes of the eighteenth century make much appeal to the sense of hearing, smell, and touch. (Tuan 1974: 140)

The parts of the park that I sought out as a child were above all those of Røhne's naturalistic garden designs that also invited impressions of 'a rich unfocused ambiance' that appealed to each and every one of my senses.

Sensuously speaking, the green lung of the Frogner Park was an Eden-like environment for a boy like me. Long before I knew the names of the many plants and animals that inhabited the park, I was endlessly taken up with them. The name of Frogner may originate from the Old Norse word *fraun*, which means 'fertile ground' – and, however highly cultivated, this was a lush environment with its wide lawns, two ponds, brook and its several thousand trees that included a few maples, elms and linden from the 1700s (a few of them more than five meters in circumference), not to speak of the dense jungle of smaller trees and bushes along the Frogner creek below the second pond, and the bush-covered slopes beneath the Monolith that offered, in their bowers, secret hiding places both for me and the sparrows. In the summer, these park environments were filled with the sounds of many different types of animals: the insistent buzzing of insects, the constant chatter of various passerine birds, the angry cries of gulls, the hissing of swans, the hoarse quacking of ducks and the short-tempered barking of people's dogs that felt the pull of ancient hunter instincts.

And what wantonness there was in this green lung! Lust lived out its life in the trees, on the lawns and around the two ponds. It was notable in the young *homo sapiens* as well who, during spring and summer, unfolded their blankets and shed most of their clothes on the park lawns, not merely to tan their winter-white skin, but to lie there necking. For me as a small boy, however, it was more fun to look at the animals that seemed, to put it mildly, much more alive: gulls on top of each other,

---

6 [Norwegian original:] 'En bør ha klart for seg at terrenget på Frogner nærmest må bli at karakterisere som ideelt, idet der fra naturens side ikke er knuslet i retning av variation, tvertimot. Terrenget er rikt utstyrt med åbne flader, høidedrag og dalsenkninger som riktig utnyttet yder det bedste grundlag for et parkanlæg av rang'.

screeching and flapping their wings; feather-shiny, persistent, amorously cooing he-pigeons who would not leave the she-pigeons be; the mallard drakes that pursued the females and tried to hold on to their feathered necks with their beaks; and the dogs that, on rare occasions, got away from their owners and pumped away in a style that, however doggy, was rumoured to resemble (according to what I had heard) that of human beings. And a little later in the summer came the busy time of feeding, when the birds noisily hunted for food for their young throughout the park: the swallows cheeped and swooped so fast that the insects popped loudly into their beaks, the finches chirped and cracked seeds, the warblers sang and nicked flies, the thrushes chattered and pulled earthworms out of the lawns and the crows and magpies tried to outcry the gulls and the ducks in their battle over dry pieces of old bread.

It was in the Frogner Park that I acquired my awe for the living world.

This green pocket in the city of Oslo also led me directly to the woods and lakes north and west of Oslo, whose first hills can in fact be viewed from the Frogner Park itself, which in turn induced me to pursue outdoor life in the mountains and moors of various locations, not least in northernmost Norway. But that, however, is another story. My main point here is not so much that urban green spaces may lead us to seek and enjoy outdoor life in more rugged environments, but that they help create in us a familiarity with, and appreciation of, the natural life of which we are all a part.

The point of the tale I have been trying to tell is this: the Frogner Park of my childhood was the natural, biodiverse crucible in which my outlook was formed before I, in a sense, knew it. The first part of a book by Jay Appleton, *How I Made the World: Shaping a View of Landscape*, is entitled, 'The World Makes Me' (1994: 1), which in my view is a very fitting way of formulating our relationship to the world in early childhood, before we consciously and intellectually begin to actively give shape to our surroundings. From my childhood wonder and fascination emerged however – by necessity as it were – an increasing interest and reverence for the world of nature, which in turn have developed into my mature, complex environmental consciousness that involves socio-political as well as ethical beliefs. My conviction that we all ideationally construct our natural environment is tempered by my respect for what I regard as its innate relations. The awe engendered by my green childhood playpen may thus be argued to have been the key to, and the driving force of, my subsequent identity development. My sense in adulthood of being deeply connected to non-human life is a direct result of having grown up, in my most susceptible years, by this variegated green space in the western part of metropolitan Oslo that is called the Frogner Park.

Some Scholarly Theories about the Role of Green Environments

Let me – for corroboration, as it were – look to some scientific arguments that have to do with the role that nature plays in our psychological make-up no matter how urban we are, which may at the same time help explain why greenways are of such paramount importance regardless of where we live. This section will first refer to theories of our genetic and biological predisposition to respond positively towards the kind of green environments that parks often embody; second it will briefly allude to some socio-cultural and psychological studies that point to the therapeutic value of green spaces and, third, it will summarily discuss why it is essential to be exposed to green spaces at our most impressionable age; that of childhood.

In his book *Biophilia*, the biologist Edward O. Wilson has argued that human beings feel innately drawn to certain types of landscapes. Certain types of terrain appeal strongly to us because we, like all animals, have been conditioned over countless generations to prefer habitats that made our survival easier than elsewhere. According to Wilson, human beings are unconsciously drawn towards certain ideal surroundings, 'no less relentlessly than flycatchers and deer mice', and Wilson's prototypical habitat example is the African savannah: 'For most of two million years human beings lived on the savannahs of Africa, and subsequently those of Europe and Asia, vast, park-like grasslands dotted by groves and scattered trees' (Wilson 1984: 109). Savannahs, in Wilson's view, were preferred because they provided an abundance of plant and animal food, offered a clear view for the early detection of dangerous animals and enemies and contained ridges and promontories that made distance surveillance possible, while at the same time providing protection at night in overhangs and caves (Wilson 1984: 110; see also Appleton 1975, Chapter 4). Since their survival has always depended on it, human beings have also needed the close proximity of water and are therefore demonstrably attracted to scenes of running water or lakes. It is striking, not least in urban environments, how often human beings have felt a need to recreate such optimal environments in their designs of gardens and parks. In my opinion, a site such as the Frogner Park with its creek, ponds, open grasslands, hillocks and its trees provides a perfect example of a landscape design of the prototypical 'savannah' we are emotionally predisposed to be attracted to and therefore enjoy. It certainly served to strengthen whatever love of the natural world was already innate in me.

Although Wilson's biophilia hypothesis has been criticized for being speculative, a great many social and behavioural-genetic studies demonstrate fairly conclusively that *biophobia* (the fear for instance of snakes and spiders, and of heights and closed spaces) is in part biologically encoded in human beings (confer Ulrich 1993: 76–86). It would therefore seem very plausible that biophilia, too, is partly genetically engendered, although research that may demonstrate this is more difficult to carry out and is still at an early stage (Ulrich 1993: 88). A great many scientific studies of people's affective reactions to various types of landscapes and cityscapes have, however, 'provided a pattern of circumstantial yet persuasive support for the genetic aspect of the biophilia hypothesis' (Ulrich 1993: 94).

Whether genetically encoded or not, biophilia, defined as people's strong preferences for certain types of landscapes, has been amply substantiated. In their study *The Experience of Nature* from 1989, Rachel Kaplan and Stephen Kaplan convincingly demonstrated that people tend to prefer the views of natural scenery of water and vegetation to the urban views of glass and concrete. In his above-mentioned article on biophilia from 1993, Roger Ulrich refers for instance to findings that demonstrate that people prefer urban parks with savannah-like characteristics (Ulrich 1993: 91, 96); that scenes with water enhance their favourable responses (Ulrich 1993: 91–2); that they are visually more attracted to urban scenes with trees and other forms of natural vegetation than to urban settings lacking nature (Ulrich 1993: 96); and that such 'similarities in responses to natural scenes usually far outweigh the differences across individuals, groups, and diverse European, North American, and Asian cultures' (Ulrich 1993: 93). In addition, there is considerable research that demonstrates the therapeutic effect on the human psyche of walking in nature or simply being exposed to natural vistas. Scientists like for instance Rachel and Stephen Kaplan (1989), Burns (1998), Lundberg (1998) and Bodin and Hartig (2003) have shown that recovery from stress, relief from anxiety or recuperation from illness are in fact facilitated through the viewing of, or participation in, natural settings (see also Ulrich 1993: 100–9 and Thompson 2008: 112).

The general findings of such studies are probably well known to a degree that makes it unnecessary for me to elaborate further on the function of green spaces in increasing and reinforcing our biophilia, our innate affinity to the natural world. I would like to end my childhood recollections of the Frogner Park by accentuating the importance of exposing *children* to such greenways. As adults living in an age of scientific rationalism we tend, as the psychologist Louise Chawla argues, to be sceptical of the child's 'sense of an animated, dynamically interactive universe' (Chawla 1994: 146) in which 'every part of the world appears alive, purposeful, willed' (145). Nonetheless, there is considerable reason to believe that this kind of early experiential attachment to one's surroundings has a particularly strong effect on one's formation. As Edith Cobb notes in *The Ecology of Imagination in Childhood*, 'Experience in childhood is never formal or abstract. Even the world of nature is not a "scene," or even a landscape. Nature for the child is sheer sensory experience' (Cobb 1977: 28–9). The child's perception is frequently coupled with a sense of wonder that Cobb sees as the very genesis of our knowledge (Cobb 1977: 27). It is, in my opinion, precisely the propensity of human beings at an early age for this type of immediate, sensory and imaginative identification with their natural surroundings that makes it so crucial for children to have access to them.

In an extensive British statistical study from 2008 entitled 'The Childhood Factor: Adult Visits to Green Places and the Significance of Childhood Experience', Catharine Ward Thompson et al. found for instance that 'frequency of childhood visits is a highly significant predictor of how often people visit green places in adult life' (Thompson et al. 2008: 124) and furthermore that, 'people who were frequent visitors as children are also more comfortable visiting woodlands and green places alone as adults and more likely to think green spaces can be magical places. It appears that such people have not just a physical relationship with green outdoor places but also an emotional one that influences how people feel about themselves and makes them more open to positive and elemental experiences in these places' (Thompson et al. 2008: 137–8). As Thompson notes elsewhere, 'The key significance of playful engagement with nature and open space for children is underlined by the way that it appears to resonate with people for the rest of their lives' (Thompson 2007: 36). In the nature writer Wallace Stegner's succinct formulation, 'there is a time somewhere between the ages of five and twelve' that leaves an indelible imprint on a man's mind: 'Expose a child to a particular environment at his susceptible time and he will perceive in the shapes of that environment until he dies' (*Wolf Willow*, Stegner 1962: 21).

This is precisely the role the Frogner Park in Oslo has played in my own life. It represented my childhood's most treasured environment, and the biophilia and topophilia it helped evoke deeply affected the ways in which I have viewed the world as an adult. The subtext of this article – its moral lesson, as it were – is first and foremost an environmental one, as I am sure my reader has long been aware. Giving children the opportunity of running around in green environments and making them acquainted with the living creatures around them represents an invaluable investment in the future health of our planet. The imprint of experiences of early childhood has a strong emotional constituent. When experienced in childhood nature exerts, in its overwhelming immediacy, an affective power that subsequent nurture – ideas of nature, for instance – lacks.

My childhood impressions of the nature of the Frogner Park are still, indelibly it seems, imprinted on my brain. I can close my eyes right now and perceive, in my mind's sensory recollections of the park, the roughness of the bark of a particular tree overhanging the lower pond, the cackling of the geese at its other shore, the mysterious contours of small fishes in its murky water and the smell of the Frogner brook in its spring flood. Biophilia and environmental understanding may of course be intellectually acquired later in one's lifespan, but if exposed to the exciting life of a green space at one's most impressionable age, one is hooked for life.

Bibiography

Appleton, J. 1994. *How I Made the World: Shaping a View of Landscape*. Hull: University of Hull Press.

Appleton, J. 1975. *The Experience of Landscape*. London: Wiley.

Bodin, M. and Hartig, T. 2003. Does the outdoor environment matter for psychological restoration gained through running? *Psychology of Sport and Exercise*, 4, 141–53.

Burns, G.W. 1998. *Nature-guided Therapy: Brief Integrative Strategies for Health and Well-Being.* Philadelphia, PA: Brunner/Mazel.

Chawla, L. 1994. *In the First Country of Places: Nature, Poetry, and Childhood Memory*. Albany, NY: State University of New York Press.

Cobb, E. 1977. *The Ecology of Imagination in Childhood*. London: Routledge & Kegan Paul.

Haug, L. 2010.Historien om Frognerparken ['The history of the Frogner Park'], *Aktiv i Oslo.* Available at: http://www.aktivioslo.no/frognerparkens_historie.php [accessed: 22 May 2010].

Kaplan, R and Kaplan, S. 1989. *The Experience of Nature: A Psychological Perspective*. New York: Cambridge University Press.

Lundberg, A. 1998. 'Introduction, and environmental change and human health', in *Environment and Mental Health: A Guide for Clinicians*, edited by A. Lundberg. Mahwah, NJ: Lawrence Erlbaum, 1–25.

Riksantikvaren ['The norwegian directorate for cultural heritage']. 2009. Vedlegg til fredningssak: Frognerparken' ['Attachment to a case of preservation: The Frogner Park']. Available at: http://www.riksantikvaren.no/filestore/MicrosoftWord.doc294383.pdf. 1–25 [accessed: 22 May 2010].

Røhne, M. 1967. *Oslo kommunale parker og grønnanlegg 1810–1948* ['Oslo municipal parks and greenways 1810–1948']. Oslo: Myhres Papirindustri.

Stegner, W. 1962. *Wolf Willow: A History, a Story, and a Memory of the Last Plains Frontier*. New York: Penguin, 1990.

Thompson, C.W. 2007. 'Playful nature: What makes the difference between some people going outside and others not?' in *Open Space: People Space*, edited by C.W. Thompson and P. Travlou. London: Taylor & Francis, 23–37.

Thompson, C.W., Aspinall, P. and Montarzino, A. 2008. The childhood factor: Adult visits to green places and the significance of childhood experience. *Environment and Behavior*, 40(1), 111–43.

Tuan, Y. 1974. *Topophilia: A Study of Environmental Perception, Attitudes, and Values*. New York: Columbia University Press, 1990.

Ulrich, R.S. 1993. 'Biophilia, biophobia, and natural landscapes', in *The Biophilia Hypothesis*, edited by S.R. Kellert and E.O. Wilson. Washington, DC: Island Press, 73–137.

Wilson, E.O. 1984. *Biophilia*. Cambridge, MA: Harvard University Press.

Chapter 4
Oslo's Ullevål Garden City: An Experiment in Urbanism and Landscape Design

Mark Brown and Mark Luccarelli

Introduction

Ullevål Hageby (Ullevål garden city) represents an important part of Oslo's urban history and development, and stands as an example of an early attempt at landscape urbanism flavoured with a strong component of naturalism derived from the picturesque. The garden city was conceived of as a template or model for utilizing urban planning to achieve social reform. It emerged out of the intellectual and political maelstrom of turn of the twentieth century London to become a staple of both British and American urban planning discourses. Garden cities and suburbs were built as cooperative enterprises outside of London before the Second World War, but the real appeal to planners coincided with the post-war period when the green aspects of the garden city were de-emphasized and its instrumental value as an implement of planned suburbanization became salient. By then, the term 'garden city' had been dropped in favour of the more utilitarian 'new town' which became a central feature of mid-20th century Anglo-American planning discourse (Mumford 1938, 1963, Osborn 1942, Osborn and Whittick 1977). Later, however, the new town and its garden city roots became associated with suburbanization and obtrusive planning by experts, especially in the work of Jane Jacobs (1961). This interpretation has, in turn, been contested by the American New Urbanist movement, which has also endeavoured to build relatively dense, town-like housing in suburban locations. However, many of the critiques levelled at new towns as representative of a faux-urban environment for socially segregated suburban housing are also directed against New Urbanism.

In our view, the garden city idea with its several examples in various countries encompasses a number of distinguishable characteristics. Against this fact, there has been a pronounced tendency to accept or reject the garden city as a model or even a tradition in urban planning. In this sense, the 'garden city' or 'new town' – often and erroneously used synonymously – is understood in relation to prevailing narratives of urban development and planning. Since the prevailing urban narrative turned against 'the suburb' (Jacobs 1961), garden cities have been seen in a negative light. In Norwegian planning and architectural history, there was a similar disenchantment but it came even earlier when it was understood that the garden city had failed to make an appreciable contribution to workable social policy in Norway. Consequently, the Norwegian experience with the garden city was more limited and even unhappier than in the English-speaking world. Ullevål Hageby failed in what has been understood as its primary mission of providing affordable housing for the emerging working class. It is seen as an experiment that was both costly and unsuccessful (Kjeldstadli and Myhre 1995: 213–18).[1]

In our view, the need to adapt our cities to green standards warrants giving garden cities another look. To reassess, the garden city begins by setting aside, at least for the moment, these established urban narratives by undertaking what amounts to an archaeological project: to dig out and reveal particular garden city experiments armed with a tool kit of green ideas about landscape and community. In effect, we are looking for what Per Gunnar Røe calls the 'green threads' – aspects of green form, function and culture – that lie all around the Oslo area (for one). They remain unnoticed, undiscovered because they are buried underneath the reigning narratives of urban development. Neither the public nor the relevant professional circles seem to recognize them for what they are or what they might signify.

In our investigation, we found three ways in which Ullevål Hageby and comparable English garden cities may be seen as a green thread. First of all, the density of the garden cities is amenable to both the sought-after features often associated with suburbs, for example, room for a small garden, and dense enough to support transit and other urban amenities. Second, the garden cities strongly emphasized landscape design as a feature of place-connectedness. Third, important features of landscape and social organization associated with an urban community, in particular the foundation for a strong civic sphere organized around an impressive public space, were achieved. In light of these features of landscape, architecture and social organization, we believe that it is possible to go beyond the rediscovery of Ullevål Hageby and begin to develop a new narrative of metropolitan development that focuses on regionalization, landscape design and common access.

1 Here is our English translation from a section of the Norwegian text: 'One shouldn't be surprised that it [the Ullevål Hageby building project] was too expensive. Rather, one can wonder why the project received the support it needed in the council in order to be realized. It must have been obvious from an early stage that these were not homes for working people. In addition to civic pride and a certain vanity, the middle class must have experienced a housing shortage as a consequence of the war just as the working class did' (218).

History of the Garden City Movement and the Development of Ullevål Hageby

Ebenezer Howard worked as a clerk in London and was a non-conformist. He was also, briefly, a pioneer who escaped from London to the Great Plains state of Nebraska to live in a sod hut far away from any city of note. But he returned from the wilderness to shape a new space for living on the margins of the great city; a new form of urban space that combined the advantages of town and country and also eliminated the disadvantages of both.

He explained this concept with his famous 'three magnets' drawing shown in Figure 4.10. On the one side was the town which sucked in people. It had clear advantages over the countryside for those migrants who were looking for a chance of betterment; a job, an opportunity to learn, religious freedom. But in leaving the countryside, many of these people had to risk their health; poor drinking water, poor sanitation and overcrowding. The city also herded them together in neighbourhoods organized by social class, so that communities that ought to have had their share of initiative-takers were left without leaders. These were very real quality-of-life advantages that the countryside could offer over the town, though it was clear enough that the former magnet was the stronger of the two in terms of pulling people towards it. Howard's garden city magnet would, he hoped, be even more powerful.

His book was first published in 1898 as *Tomorrow: a Peaceful Path to Real Reform*, a title, which reveals Howard's chosen role as a social reformer. His subject attracted popular interest and Howard was invited to lecture all over England. He attracted a small number of able supporters and in 1899 the Garden City Association was formed. His book was reissued in 1902 as *Garden Cities of Tomorrow*, and in 1903, just over 1,500 hectares (3,800 acres) of land were purchased in Hertfordshire about 35 miles from Central London. The new owner of the land was First Garden City Ltd and, under the direction of its newly hired planners Raymond Unwin and Barry Parker, the corporation commenced the building of an experimental town on Howard's guidelines; Letchworth garden city came into existence.

The establishment of Letchworth, however, did not, as Fredric Osborn in his preface to an edition from 1965 observes, 'lead to the widespread foundation of new towns on the same principle' (Howard 1902/1965). But there were scaled-down versions that lacked the independent industrial base and open spaces. One famous example was Hampstead Garden Suburb (founded in 1907) on which Ullevål Hageby was modelled. Like its forerunner in London, Ullevål Hageby had neither industries, nor a greenbelt. Both were located within the immediate pull of the existing cites, London and Oslo respectively. Nonetheless, they retained features of the garden city idea; namely social reform through improved housing and landscape design.

Ninety years ago, Ullevål Hageby was a cutting edge residential development on the edge of Christiania (the earlier name for Oslo). At the time of its construction, between 1916 and 1926, Christiana had a population of about 250,000 inhabitants. There was strong segregation by social class on an east-west axis, and also an equally well-defined distinction between town and countryside. Oslo had clear borders then. Urban life was the province of a small minority of Norway's total population and was characterized, on the one hand, by tenement housing for the working class, or on the other hand, by spacious, well-appointed 3–4 story apartment housing for the privileged classes. The culture was, then, truly urban in that it was pedestrian-centred and oriented around the business and cultural activities made possible by the city for those who could afford them. Sunday walks by the bourgeoisie were on Norway's most civilized street: Karl Johansgate. It was an age of high urbanism based on a German model of the 19th-century bourgeois city. Ullevål Hageby represented an alternative to these urban modes – both high street and the tenement; it combined features of landscape and garden design that created a friendlier and more relaxed way of life without sacrificing important urban features.

Today, Ullevål Hageby is a high-end residential area of west Oslo. It houses a community of well-paid professionals – an enclave of Richard Florida's creative class (2002). The area is semi-suburban (by today's standards), attractive and located in the innermost suburbs, well within the city limits and close to the city centre. It is denser than the surrounding housing and contains a vibrant town centre, both commercial and civic. On our visit, we witnessed that the long-time bakery in the town centre has become extremely popular as a café; it was packed with students from Oslo University, which is a brief walk away. In all respects, it is superior to its surroundings and has maintained an aura of urban and natural values – truly an alternative to what we think of when we say 'suburb'.

In his introduction to *Garden Cities for To-Morrow*, Lewis Mumford claims that the garden city idea has been the most profound single influence on town planning, but suggests that 'it has met the traditional misfortune of the classic: it is denounced by those who have plainly never read it and it is sometimes accepted by those who have not fully understood or unconsciously modified it' (Howard 1902/1965). As we will seek to demonstrate, the garden city idea has been misunderstood. Elements of it can, we believe, make a useful contribution to green urban discourse.

Ullevål Hageby and Landscape Design

Green cities, smart growth, controlling development in the semi-urban areas or any other planning concepts applied to the problem of urban sprawl tend to rely directly on the limits to the growth argument. As a consequence, the environmental discourse has been couched in a language of imposition of rules, restrictions and givebacks without compensation, rather than as an invitation to realize what Lewis Mumford once described as 'more exhilarating' and 'fuller human environments' (Mumford 1926: 91). Design in this sense may be understood as a supplement to regional planning, in that it suggests a set of linkages from infrastructure to ecology and aesthetics that is rarely addressed by urban and regional planning. Place identity rests on the cultivation of visual sensibility, as interpreted through cultural processes and discourses. It also requires attention to social networks which link human activities and particular spaces. Ullevål Hageby was built with great attention to landscape design, creating an aesthetic or pictorial value, intended to improve the quality of life for its residents and add to the qualities of the city as a whole by developing an identifiable neighbourhood. This is not to deny that these particular identifications of place have been attenuated, as they were subverted by mainstream discourses and practices aimed at gentrification for social elites. Our point is that attention to landscape design is a recoverable value that may be applied in a different socio-spatial context that is emerging as our future.

At one level, our approach to the study of Ullevål Hageby may be likened to that of archaeologists rediscovering a lost artefact – albeit an artefact that is now occupied; it comprises its own living community and is a resource of the urban landscape that may be put to new uses and conceptions of living. To begin with, the very fact that it is a designed landscape makes it a good laboratory for studying the morphology of suburban design as well as an excursion into the process of building landscape urbanism, even though it was created on a very different basis than current landscape urbanism projects. Our point is that while urban design has been an important part of city planning discourses, landscape design has been largely missing, partly because of the privatization of design and planning, which must perforce respond to market forces.

Contemporary European landscape urbanism is an attempt to bring concern with design back to the forefront of landscape architecture (Shannon 2006: 146) and thereby overcome the diminishment of design brought about by the triumph of a positivistic formulation of landscape ecology over the last 30–40 years (Treib 1999).[2] Landscape design has become important again as urbanists realize that the process of urban diffusion has made it difficult, and counter-productive, to identify places in architectural terms, that is, by the design and arrangement of buildings. Diffusion unto the land means that if cities are to make visual sense and if they are to develop place awareness, then incorporating design concerns into landscape is a necessary response. Contemporary landscape urbanism's common design emphasis belies the different approaches of practitioners and theorists, though there are several common threads: the importance of infrastructure to landscape design, the definition of landscape by reflecting layers of use over time, the accommodation of ecological concerns and a hybrid aesthetic sensibility based on the 'mixing of natural and cultural systems' (Weller 2006: 73).

In the case of Ullevål Hageby, there is much stronger identification with nature in landscape design, which is logical since this is the predominant Anglo-American tradition and formed the most important influence on Ullevål Hageby. This has certain drawbacks and also great strengths though it is beyond the scope of this

2 Nonetheless, the ecological dimension of landscape remains important in landscape urbanism.

chapter to develop the theoretical implications of naturalism in landscape. Our purpose here is to characterize the design principles at work and to criticize them in the context of their use. The goal is not to promote Ullevål Hageby as a model for redevelopment, but to identify its landscape principles, to show the relation of these principles to concepts of nature and space and to judge their potential importance to future urban landscape redevelopment projects in Norwegian and other contexts. These principles of landscape may be applied in different contexts and may be appropriate for different parts of the evolving green urban region. Redevelopment is also a form of new development. It is not our task as cultural historians to explain how garden city principles might be adapted to new circumstances, but simply to explore the principles apparent at Ullevål Hageby – and secondly English garden cities, within their historical-cultural contexts.

Principles of Landscape Design

Tim Beatley writes about 'biophilic cites' – cities that demonstrate a reverence for the natural landscape and a connectivity with the natural world. Oslo's surroundings – the presence of the fjord and the forested hills surrounding the city to the north – create what might be called an ecological, 'extensive and inclusive ground-plane of the city, to the "field" that accommodates buildings, roads, utilities, open spaces, neighbourhoods, and natural habitats' (Wall 1999: 233). This ground-plane impacts on recreation and to some extent contains the city's sprawl. Of course, this is due to regulations, which have in effect created a boundary line (the *markagrensa*) between the forest greenbelt and the city. But it is also due to a discourse of the forest and its cultural significance. Our concern here is rather with how landscape impacts the design of a residential area of the city – it is a concern, in other words, with a finer level of detail that impacts everyday life.

Landscape as it developed at Ullevål Hageby may be understood as a *Zusammenhang* – a sphere or a field connecting unlike aspects of experience. One of the qualities that makes Ullevål Hageby virtually unique in the Norwegian experience is the assumption that planning and cultural discourse around planning can mediate the line between private and public. In a public-oriented landscape tradition, private housing is often seen outside the landscape, which is identified as common space (Thorsen 2011). Ullevål Hageby was a cooperative enterprise and in its civic structure as well as the landscape design, there was an attempt to infuse the private sphere with values of civic-mindedness and to rely on the spirit of a local community to define the sense of the public. These are important qualities in the development of discourses necessary to green urbanism.

The second sense of landscape as a *Zusammenhang* pertains to mediating the culture-to-nature divide. Landscape as an idea has been criticized for its pictorial and abstract qualities (Cosgrove 1998), but it also has a long history as a pictorial and scientific means for tracing the impact of human activities on nature, and of conceiving the physical geographies and ecological functions that underlie human settlement (Jackson 1999). Bridging nature and culture is also a quality of landscape architecture. Indeed, the construction of any designed landscape entails a conceptual understanding of space and of nature (sees Tables 4.1 and 4.2). The two most important aspects of all landscape design, as we understand it, are first the capacity to have design reflect natural principles, and second the ability to incorporate, without conflation, social space and natural features. The means of defining space and the conceptual understanding of landscape as design vary; below in Tables 4.1 and 4.2 we have applied these principles to reading Ullevål Hageby.

Table 4.1 Nature as source of landscape design at Ullevål Hageby

Concepts of Nature	Corresponding Principles of Landscape Design
(1) Ecology and cellular biology – ecology is the interaction of organisms, and each organism has a nucleus and cell walls.	(1) Cities are landscapes composed of may different places each of which requires boundaries and centres.
(2) Variation and spontaneity underlies function, uniting ecology and design	(2) Variation around a theme, intricacy and curvilinear forms (the picturesque).
(3) Cultivation – the preparation, use and caring for the land, a prime aspect of human nature (man the cultivator)	(3) Landscape embedded with gardens entail connectivity with place.

Table 4.2 Space as source of landscape design at Ullevål Hageby

Concepts of Space	Corresponding Principles of Landscape Organization
(1) Centrality is necessary as point of focus of social life.	(1) Central space, its uses, its relation to natural space and its surroundings
(2) Diffusion is necessary as population overtaxes capacity of a city and desirable to provide green spaces.	(2) Variation in housing types in relation to location.

Our study arises out of an appreciation for the achievement of the garden city, even though, unlike advocates of New Urbanism, we are not presenting it as a model for the future. We believe that examining the visual order of the garden city and how it was created is an important key to the creation of new cityscapes that are not simply orderly, but speak of the values of nature and landscape as they have evolved in Western culture over the past 200 years. We have arrived at five basic landscape/design principles based on concepts of nature and space: first, that ecology means the interaction of organisms; second, that although the metaphor of the city as a single body has been used before, empirical observation tells us that a city is far too large to be a single organism; third, that ecology suggests a mosaic of distinct areas, each of which has its own identity and boundary;[3] fourth, that ecology presupposes a fundamental unity achieved through variation, sometimes called in design, 'variation around a theme'; and fifth, that built environments (including architecture and landscape design) should draw on the larger cultural landscape, that is the vernacular architecture and landscape forms of a region. While a discussion of principles one and two are beyond the scope of this essay, in the next three sections we illustrate how principles three, four and five have been realized using Ullevål Hageby and English garden city and new town examples.

Landscape Expresses Boundaries and Centres

The third principle is illustrated at Ullevål Hageby both in the original design and, to a lesser extent, in its current form. In the original winning design for the

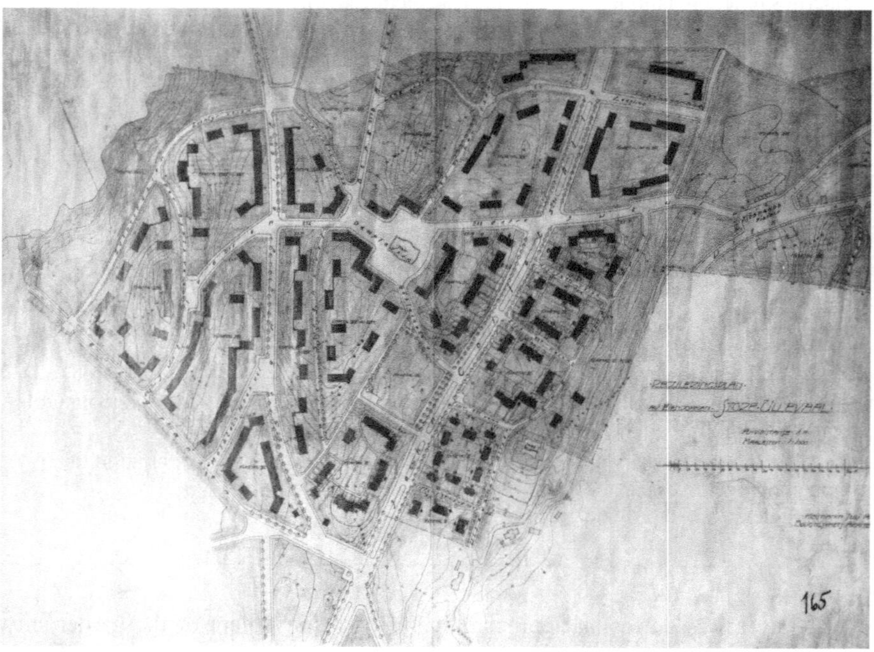

Figure 4.1 1913 plan for Ullevål Hageby

Source: Map by permission of the Oslo City Archives.

3 This concept is related to 'urban ecology' as practiced by Burgess and Hoyt of the Chicago School of Sociology.

development of a largely open 43 hectare (106 acre) parcel of land, the architect Oscar Hoff utilized the features of the natural landscape to accentuate boundaries. One glance at Figure 4.1 is also sufficient to register the focal centre Damplassen that he created around the site of a small pond, turning it into an ornamental pool.

The northwestern fringe slopes down away from the centre; Hoff's plan has left it free of housing. Similarly, the area running from just east of the southern apex in a north-easterly direction slopes down, away from the centre and has been left free of development. When the area was first developed in the 1920s, the vantage point from both south east and north west would have looked upward towards the new garden suburb, which was taking shape on the higher ground. These areas were not intended to be part of the landscape gardening and what remains today is still, by the standards of the Oslo parks department, rough ground. Although it does not

Figure 4.2 View of Hampstead Garden Suburb taken facing northwards from Hampstead Heath. In the centre, the spire of St. Jude's church

Source: Photograph by Mark Luccarelli 2010.

have such imposing buildings at its centre as its English cousin Hampstead Garden Suburb, the vista of Ullevål Hageby when approached from the south east in the

1930s would not have been dissimilar to the one, which is still available today in London (Figure 4.2).

As Oslo has expanded outwards and begun to absorb Ullevål Hageby into its urban structure, these natural boundaries have become both attenuated, and at the same time even more significant in preserving a distinct physical identity. The green space of the north-western fringe, in particular, was compromised by the construction of a metro line, opened in 1934, which runs right along the middle of the strip. And both here and at other places around the site, one presumes that commercial development pressures have diminished the boundary's integrity. However, the experience of crossing boundaries and entering into a distinct place still remains today as one

Figure 4.3 Woodland on Norton Common, Letchworth garden city

Source: Mark Luccarelli, 2010.

walks around; this sensibility is greatly augmented by a distinctive architecture that strikes us as neither specifically urban nor suburban.

Boundaries are also an important aspect of the urban landscape in the English garden city projects. On the western edge of Letchworth garden city, not more than ten minutes' walk from the town centre, is Norton Common, a very large area of wood and parkland in which birdsong comes to the auditory fore and automobile traffic declines to a distant hum. Norton Common is not wilderness. But its woodland area is managed for the benefit of the wildlife (Figure 4.3).

Hampstead Garden Suburb is not as fortunate as Letchworth in being able to

Figure 4.4 The Great Wall (left) and the stream in Northway Gardens (right), both features of Hampstead Garden Suburb

Source: Photographs by Mark Luccarelli, 2010.

maintain an enormous area of woodland on its boundaries. However, Raymond Unwin, who moved down to London from Letchworth in order to take on the task of developing the suburb, did fashion boundaries to provide the demarcation that contributes to identity building. The Great Wall, a red brick construction along the southern border is shown, covered in ivy, in the left of Figure 4.4. The right-hand picture is a stream, which runs through the centre of a narrow strip of gardens separating the main area of the suburb from the Falloden Way, a busy arterial road leading into the centre of London.

The centre of Ullevål Hageby is Damplassen, (the place of the pond). Hoff had to accept a design constraint when he began work on the 43 hectares of land. He was forced to draw in a new road, John Collets Allé, along the course of a large pipeline carrying fresh water into the city. In parallel with the development of the suburb, a road running north –south, Sognsveien, has acquired greater significance than perhaps originally envisaged as an important arterial road for traffic entering and leaving Oslo city centre. As a consequence, Damplassen does not enjoy the status of being a centre in the experience of a car driver. However, this is probably considered to be an unmitigated advantage by all the pedestrian users of Ullevål

Figure 4.5 Damplassen, the centre of Ullevål Hageby (pond not yet filled with water after emptying for Norwegian winter)

Source: Mark Luccarelli, 2010.

Hageby. An approach to the centre can be made from all points of the compass and the architecture changes to a more imposing Georgian style. As one enters, the vista opens out into a large area with the ornamental pool at its centre, flanked at an appropriate distance by the grander buildings (Figure 4.5).

The final point that we believe is very important in making for a vital centre is to provide people with reasons to go there and to congregate. This point was brought home to us most vividly in the comparison of our experiences in Ullevål Hageby and Hampstead Garden Suburb. In the latter, the absence of shops at the centre and its domination by two great churches (a legacy of Dame Henrietta Barnet's guidelines for the development and management of the estate), has led to there being virtually no people present in the centre when we were there. At Ullevål Hageby, however, we enjoyed a coffee and pastry in a well-known bakery that buzzed with conversation and looked out on a centre with several shops, which, while not exactly bustling with activity on a chilly winter's day, certainly had more life than similar residential areas of Oslo.

Landscape Expresses Variation and Difference around a Theme

Raymond Isaacs has reviewed literature from the fields of psychology, geography and urban design to examine the 'picturesque' urban environment (Isaacs 2000). The picturesque is both a cultural motif, with multiple artistic and intellectual origins that emerged in England in the 18th century (Gilpin 1782), and a set of principles for designing landscape parks as practiced by the American Frederick

Law Olmsted (Central Park was laid out in the 1850s). At the same time, the picturesque may be understood as attributes of the environs (surroundings) when they are perceived as both delightful and stimulating. In such cases, the environment is said to contribute to the sensibility of environmentality – our individual and common awareness of and attraction to our surroundings (Isaacs 2000). One essential aesthetic attribute of the picturesque is the appeal to both complexity and order; the sense that variation and difference find a necessary thematic unity. This picturesque motif may be understood conceptually in terms of the ecological and the natural.

As an aesthetic, picturesque reflects the artistic idea of the natural as mimetic suggested by the English Pre-Raphaelite movement. Immersed in the study of nature from which they gathered attentiveness to natural details, and inspired by the late medieval/early Renaissance Italian painters, the Pre-Raphaelites painted with a strong sense of colour and directness in theme (des Cars 2000). The particular relation between variation and theme we have noted is however more directly suggestive of the development of the idea of ecology in 20th century environmental thought. Ecology emerged as a powerful metaphor of the natural, expressing the capacity of nature to encompass spontaneity and development, or freedom and evolution, as an interplay between order and variation (Worster 1985). We are suggesting that biological principle is very much in evidence in the best of garden city landscape design.

Equally important has been the fact that picturesque principles, often seen as rural qualities or attributes of wilderness, actually have a long history of being applied to the urban world. John Ruskin's valorization of the gothic in both architecture and landscape is related to picturesque, and his work on Venice (1853) suggests that the rough and irregular senses of the natural that lay at the origin of gothic, created a foundation for what we find the most endearing (picturesque) qualities of Renaissance cities. Lewis Mumford picked up this idea in his treatment of Venice as an example of a natural regional city (1963). More recently, research has suggested that the love of the picturesque survives in the unexamined taste of many members of the public, which may explain the appeal of old towns with their narrow twisting streets and common spaces (Isaacs 2000).[4] At Ullevål Hageby and its English counterparts, the order, which our aesthetic sensibilities require, is provided by the architecture and landscape architecture. The four pictures in Figure 4.6 are all of residential accommodation in Ullevål Hageby.

There is an architectural theme that is present in all the buildings. The building materials are of stucco-faced brick. All of the buildings have a pitched roof, the

4 An alternate constructivist explanation as formulated by Bourdieu (1984) would emphasize the power of cultural capital to shape tastes and create fashion in line with the need to maintain social stratification. By contrast, while social stratification can operate through cultural discourses of taste, we argue that aspects of taste have near universal appeal because they respond to felt wants and because they correspond to established and widely shared cultural processes.

Figure 4.6 Residential accommodation in Ullevål Hageby

Source: Photographs by Mark Luccarelli, 2010.

roofing is carried out with the same red tiles and the windows are the same. From a landscaping point of view, all the buildings have garden around them and trees are in evidence in each picture.

As observers, we feel a balance between order and variation. Architectural details are an important part of a landscape that stimulates the senses and excites curiosity. The pitch of the roofs varies from building to building and on the same building. Dormer windows are cut into them on some houses but not others, and the technique varies too. On some of the buildings, a wooden facing replaces the stuccowork on the first floor. The landscaping varies too although it is not clear from these photographs. Some buildings are placed near to the road with larger gardens at the rear, while others are set back with large front gardens. Some apartment units are constructed with a central courtyard and, if one examines chief architect Oscar Hoff's plan shown in Figure 4.1, it is clear that the arrangement of the buildings into subsections provides for a varied visual experience as one walks around the suburb.

In Figure 4.7 we have provided examples of some of the smaller details that draw our attention to the specific topography of the site, stimulate the senses, and excite curiosity. The street, top left, curves around to the left. This leads us to wonder what lies beyond the curve – one of the principal techniques of picturesque. Note too, that the row of terraced houses is built in a curve rather than being divided up into a series of straight, discrete sections, and that there is a setback in the row which disturbs the curvature. At the right-hand end of the terrace is an archway which creates a boundary between the public street and the more private courtyard that is shared by the residents. Demarcations points are a regular feature of Ullevål Hageby's architecture. Note the detailed brickwork around the arch, top right, which leads one through to the interior courtyard of one of the imposing residential buildings in *Damplassen.* On another curved street, bottom left, the landscapers have preserved a tree that stands in the roadway, even though its positioning interferes with traffic flow. In the picture at the bottom right of Figure 4.7, a driveway leads round to the rear of a building but a narrow path leading to some unknown destination is visible at the end of the drive and invites us to explore further.

Architecture as Landscape and Memory

The fifth landscape/design principle can be accomplished by incorporating the vernacular and the landscape of the countryside into the urban fabric. Architecture can refer to the cultural landscape, both vernacular and formal. We can illustrate the principle using examples from both Letchworth and Hampstead Garden Suburb. Camillo Sitte's *Die Städtebau nach der künstlerischen Grundsätzen* (1889) praised the informal layout of medieval towns. According to Peter Hall, Raymond Unwin had similar inclinations and read Sitte, though not until after designing Letchworth (hence the formalist aspects to much of its design). Later Unwin published *Town Planning in Practice* (1909), which reflects Sitte's idea that the achievement

Figure 4.7 Stimulating our interest – variations and details in Ullevål Hageby

Source: Mark Luccarelli, 2010.

of medieval informality required close attention to the relation of buildings to particular spaces. There must be strong attention to details, which Unwin derived by making line drawings of medieval English, French and German towns and villages (Hall 2002: 103–4). At the same time, Unwin and his chief collaborator Barry Parker,[5] have managed to design structures inspired by their interpretation of the cultural landscape, especially but not exclusively English. In Figure 4.8, we have juxtaposed pictures taken in our two English garden city examples on the left with photographs of structures embedded in the cultural landscape.

In the top pairing of Figure 4.8 is the house that was constructed in Letchworth for Barry Parker, the architect who is most closely associated with the city's development and which now houses the small museum of the city. The form of this house mirrors very closely that of the building on the right, Prior's Hall Barn in Widdington, Essex, which is described on the website as 'one of the finest surviving medieval barns in eastern England' (Edwards 2011). In the middle pairing is a picture of Meadway Court in Hampstead Garden Suburb. Although it is an apartment building, its form immediately creates an ecclesiastical connection in the mind. On the right is

5 Edwin Lutyens was an additional contributor. All three were influenced by the style of the Arts and Crafts movement.

Figure 4.8 Garden city architecture (left-hand pictures) resonates with architecture found in the wider English and continental cultural landscape (right-hand pictures)

Source: Photographs i, iii and v by Mark Luccarelli, ii by Robert Edwards, iv by Iain Robertson and vi by Mads Angelbo.

a picture of the cloisters of Norwich Cathedral. The bottom pairing in Figure 4.8 shows on the left a section of the Great Wall in Hampstead Garden Suburb. What is, in fact, one of a series of garden sheds that are set into the wall at regular intervals, bears an unmistakable resemblance to a guard turret on a medieval city wall. To the right we see the reconstructed city wall of Brasov in Rumania. This was a city that

Figure 4.9 Garden city landscape architecture (left-hand picture) resonates with a natural features of the vernacular landscapes (right-hand picture)

Source: Photographs i by Mark Luccarelli and ii by Immanuel Giel.

was dominated by German merchants and guilds. They built Brasov in the German medieval style to which they were accustomed and which Unwin greatly admired.

We use a similar juxtaposition of photographs in Figure 4.9 in order to illustrate ways in which the landscaping of the two English examples corresponds with an English vernacular countryside. The picture on the right shows part of the woodland of Norton Common on the edge of Letchworth featuring the essential gnarled old tree. The photograph on the left is taken from the entrance to a small park in Letchworth. The pathway through the trees curves into what appears to be a great distance, a typical effect of picturesque landscape gardening.

With these two sets of pairings, we have illustrated the different ways in which the creators of the English garden cities attempted to evoke the past and connect a new city with landscape memory. The same can be said of Ullevål Hageby, except that the landscape and architecture upon which it draws is more continental than Norwegian. As observers of Ullevål Hageby today, we do not experience continuity with a Norwegian vernacular. The architectural style that Hoff used owes more to the work of the English architects than to any native Norwegian sources. It is safe to say that he drew heavily on the work of Unwin and Parker. Ullevål Hageby's shortcomings in this regard may have two causes. First, the tradition of bringing landscape into the proximity of a city or densely populated area – think of Central Park in New York – may not have made cultural sense to the Norwegians, for whom landscape is carved out of the forest rather than preserved as a pastoral countryside (Norberg-Schultz 1996). Second, the Norwegian natural landscape is so rugged and forested that it is less amenable to creation as a park than a countryside landscape is – though it would be interesting to see a landscape architect's attempt to illustrate the feel of the Norwegian landscape in a small park of a few acres.

Our larger point is that the capacity of landscape design to evoke an environmental sensibility is evident in the garden city. The garden city is most significant as an example when it generates a vital, interesting landscape on the basis of the familiar and the provocative. To the extent that it did this it could generate a deeper sense of the significance of environs, a sensibility that in today's world could be applied to developing support for green practices and policies. There are really two separable aspects to this. One is that landscape design in general generates biophilic feelings. These designs are not place-specific in the sense that they could be experienced anywhere, but arguably they are capable of generating a general sense of environmental awareness. The second point is that the garden city reflects on specific types of environmental awareness related to the immediate community, on the one hand, and the larger cultural landscape on the other. These are place identities that encompass locality, nation and civilization – but express them for the time in an interesting way. The garden city landscape strengthens a sense of local place, but also refers to the national landscape and by allusion to earlier Western landscape traditions, that is, to lost mediaeval townscapes and to landscapes that reflect the building and farming practices of peoples (English and Scandinavian) located on the periphery of European urbanism. These themes introduce a sense of variability as exotic and therefore flirt with reducing picturesque to a trope. The power of the garden city as a landscape, however, rests on a powerful combination of immediate experience and cultural symbol. It is not within the purview of this article to discuss to what extent, indeed if at all, such conceptions of landscape symbolism are relevant today, or in what way they were effective then. Our point is more general; landscapes are designed and can (if the designer so wishes) stimulate the natural aesthetic and biophilic sensibility, and develop senses of the native and the exotic and their interdependence.

Regional Planning and the Garden City's Ideological Landscape

We have discussed the garden city in reference to landscape design, but it also constitutes a contested ideological landscape that has been understood within the contexts of urban development and urban planning discourse. One must admit that at first glance there seems to be little to contest. To take the garden city seriously in the context of sustainable urbanism is at best paradoxical and at worse absurd. Urbanism means, first and foremost, that people live in cities, rather than abandoning them at 5pm. It is important that this includes people of all social classes, though arguably the most important model for constructing attractive urban areas has been generated by the civil society, the market and the state: that is, the processes of discovering run-down neighbourhoods by socially marginalized groups, such as bohemians or gays, followed by middle-class invasion and gentrification.[6] Policies that build on this success will permit areas with a strong residential character, mixing housing with commercial functions within walking distance. Garden suburbs, by contrast, have been understood within the context of incentive-driven suburbanization of the post-war period, and in relation to a mid-20th-century urban planning discourse that ignored city centres in favour of metropolitan diffusion. As such, in the quest for sustainable urban forms, the prevailing view today is that there is very little use for the garden city.

Our response to this is twofold. First, garden city planning's history should not be seen as synonymous with suburbanization. There is an important distinction to be made between diffusion and multi-nodal or poly-nucleated urban development. Second, and of more importance in regard to Ullevål Hageby in particular, the garden city was the first awkward step towards the idea of an urban landscape. Today landscape urbanism, the idea that landscape and not architecture is the measure through which the contemporary city is understood, is an established concept (Waldheim 2006), but the development of this idea surely owes much to the example of the garden city.

The garden city reflected both a designed landscape suggestive of landscape urbanism and an ideological one. We consider the more complex landscape design issues below. In the context of the development of regional planning, the ideology of the natural landscape was very important and the garden city was a focal point for these developments. Garden cities attracted interest because they suggested that a new post-urban landscape could reflect the goals of Utopian socialism by encapsulating the pastoral values of simplicity, naturalness, peacefulness and a strongly communitarian social ethic.

There was nothing new about the pastoral landscape, of course. Exurban landscapes had since the early Renaissance been imbued with pastoral values and were understood as retreats from the city, its public culture and political responsibilities: Palladio-designed neo-classical villas for gentlemen who wished to enjoy noble and edifying surroundings with beautiful gardens and pleasant country walks, far away from the counting house or political strife of the city. This reflected the emergence of a new merchant-aristocratic class made possible by the particular economic and social developments that characterized Italian city-states at the end of the Middle Ages. There were several factors. The cities had survived the collapse of Rome, limiting the influence of feudalism and its pattern of isolated rural life. The establishment of trade quickly revived them and, finally, the aristocratic class had always participated in urban life, and had quickly intermarried with the rising mercantile class (Braudel 1949). The pastoral countryside was a landscape born to serve as a retreat for this new class, a motif that extends from the Renaissance to the present (Duncan and Duncan 2004). Still, the aristocratic-mercantile class managed to distinguish itself from others by virtue of separation from the city for periods of time. Rural pursuit of the cultivated countryside living through architecture, painting,

6 State-directed gentrification projects have become important generators of economic activity. Bjørvika in Oslo is an example.

and horticulture was highly valued for intrinsic reasons; the countryside was thought to reflect purity, clarity and cultivation – the foundations of culture. At the same time as the countryside view acquired on a walk beyond the town walls, the *paessaggio*, gave way to the countryside estate, it became obvious that landscape developed a social value; it was a scarce commodity that could not be enjoyed equally by all. As trade restored cities throughout Europe, this Italian pattern established itself along similar lines in England. In both places, love of countryside and the establishment of the country seat quickly became a mark of the upper and middle classes (Cosgrove 1998) and cultivation of its virtues was seen as a prime attribute of European culture (Emerson 1892).[7]

In short, landscape became what eco-critics call a pastoral trope (Alpers 1996) that was juxtaposed to urban life. In some formulations of pastoral it represented the antithesis of urban life as famously championed by Jane Jacobs (1961). The association of pastoral values in the ideological landscape of the aristocratic countryside began to change in the period of industrialization and hyper-urbanization. In the work of the art critic John Ruskin, landscape became an agent of social reform; the aesthetics of the countryside was seen as a remedy to the social and physical ills of heavy industrialization. The parks movement in America, led by Frederick Law Olmsted, the designer of Central Park which opened in 1858, argued that public access to landscape in the heart of the city was the greatest remedy for the ills of the commercial city. In this era, the qualities of a pastoral landscape began to change. It became associated with a new aesthetic of naturalness associated with the English garden and the picturesque.

In 1890, William Morris published *News from Nowhere*, a science fictional account of a future world that prophesizes the achievement of a socialist utopia without private property or compulsory labour and without great cities. Work has become pleasure and London has been abandoned. Gone with it are all of the atavisms associated with the Victorian city. What has taken its place looks very much like pre-industrial England; a village and rural society that are nonetheless socially far advanced from those conditions which prevailed under the regime of liberal industrial capitalism. The demise of the great city was possible by decentralizing the population into a broad region – in effect a process of 'socialist suburbanization'. It was a pastoral cultural vision in the sense of both contemplating a retreat from the complexities of urban life and embracing the simplicity and honesty of the countryside. Morris was both a socialist and a founder of the arts and crafts movement that favoured a return to craft production. In his formulation, pastoral landscape became an important marker for the achievement of social peace and a necessary agent in achieving a Utopian political vision. By presenting a Utopian arcadia, Morris thus gave a very different reading of landscape than that associated with aristocratic retreat (Gifford 1999: 36). The landscape and townscape generated a sense of place that seemed essential to social reform – an early example of place-centred social thinking. Commitment to participation in

7 A different pattern held sway in France with its emphasis on court life.

shaping one's physical and social environment is a necessary part of both cultural self-determination and appreciation for and concern with nature. Landscape, then, became a common thread in the social and aesthetic concerns of Morris's vision.

Eight years after the appearance of Morris's book, Howard published *Garden Cities of Tomorrow*. Howard absorbed Morris's formulation, linking the motifs of pastoral retreat to appreciation of traditional English countryside, small-scale production, village or town centres, and common ownership. What changed was the political context. Howard argued, in effect, that social and cultural change could emerge within the contexts of existing social relations and metropolitan regional structure. Thus, the original title of *Garden Cities of Tomorrow* was *A Peaceful Path for Reform*. The methods were, in typical Anglo-American fashion of that time, based on voluntary associations operating through and around market mechanisms, rather than through state planning. The social benefits of home life would be available to the masses by making housing available as affordable cooperative housing, and this would be augmented by a new found reverence for landscape and the wholesome joys of country living.[8] The garden city proposed to reform society through reorganizing spatially, though the spatial reorganization would be integrated with social change. Howard's original proposal for the garden city would include parks and agricultural land consisting of a greenbelt. At the same time, the garden city resembled continental European planning for satellite towns to be built at the metropolitan edge. It would be a self-sufficient town attractive to industry, which could be fitted unobtrusively into the design of the town. In Howard's interpretation of the countryside, the presence of a visible landscape and pleasant town settings was a path to social peace, and mixing labour with the land, a useful practice in building community ties.

Its regional character is apparent in Howard's original motto for the garden city: 'the perfect marriage of city and country'. Unlike a suburb, the garden city was meant to be its own town and to be productive, incorporating the Georgian pastoral ideal of the dignity of simple labour with a practical scheme for attracting non-polluting industries. Thus, it was not really a vision of suburbanization, but rather an attempt to reform the impulse towards suburbanization by transforming it into a third force that would coexist with the existing city and countryside. This is clearly implied in Howard's 'Three Magnets' diagram (Figure 4.10). The diagram shows the garden city as the perfect balance between countryside and the city, but it also shows that the garden city is but one of three poles.

It is assumed, in other words, that the existing polarity of city and countryside would remain. The garden city, in effect, was an exercise in landscape-urbanism that if applied generally would strengthen existing forms. This implied a regional approach, for only in a regional context could the three magnets coexist, a point

8 Letchworth, the first English garden city, was intended to eliminate private ownership. It was a housing cooperative run by a board providing space for housing on 99-year leases. Its first housing exhibit was a cheap cottage meant to be affordable to the working class.

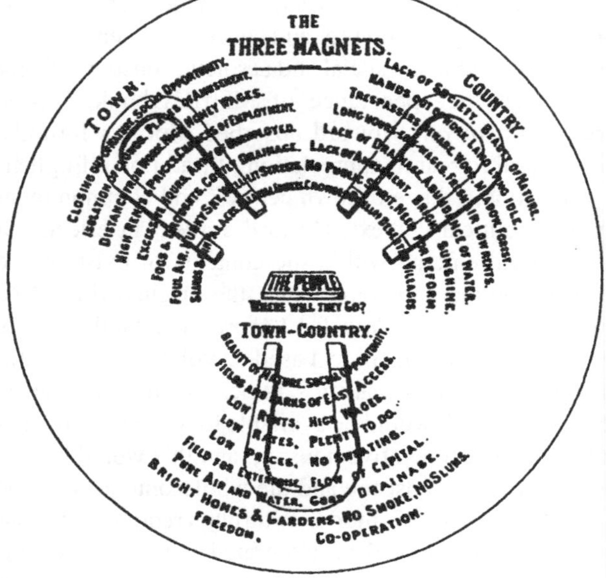

Figure 4.10 The Three Magnets diagram

strengthened by a diagram showing the network of garden cities connected by rail lines. It was only in this context that the garden city could be seen as having a general reformative effect on the metropolitan urban complex.

Garden City versus New Town: Commentary on Post-war Modernist Planning Discourse

After the war, modernist planning discourse accepted urban diffusion as inevitable and raised few questions about the impact of migration out of urban places. The future was defined in terms of new kinds of lower density settlements to be built at the urban edge. Suburban housing, often free-standing structures with individual gardens, was typical of Anglo-American experience dating back to the 18th century (Fishman 1988). Garden cities began a discourse that culminated in new town planning, which was widely seen as an alternative to minimally planned housing estates.

Planning discourse, whether in praise or criticism, has always tended to see the new town as an extension of the garden city (Abercrombie 1945, Jacobs 1969). However, one of the first things we noticed in researching garden cities outside of London was how strikingly different they were from the new towns. Post-war British and American planning discourses were strongly influenced by continental modernism, particularly the emphasis on creating a landscape modernism in which the new city would be shaped by perception of speed and motion as worked out in

Le Corbusier's *La ville radieuse* (The Radiant City) of 1933. The traditional centre was judged obsolete – built for the speed of the horse-drawn carriage. The new city would reflect the perspective garnered from the highway. Indeed the highway, the landscape park and the high-rise building were welded together into a coherent whole; the creation of the modernist city that reflected the search for a new social order. The highway complex was matched by the new social sites of the city: model factories, community centres and nursery schools that were in line with the Fordist model of accommodation between big capital and labour seen in all advanced societies in one form or another (Hall 1999: 937). In some respects, this is reflected in earlier planning discourses. The City Efficient thinking that dominated American planning after World War I had no grand vision of a new city, but it was just as concerned with adopting the city to the motif of speed and movement in an age of motors. Efficiency in circulation and structural differentiation were its chief goals. The ideal city moved traffic quickly and designated separate areas for different urban functions (Luccarelli 1995). What changed after World War II, was both the size and scope of modernization and the triumph of accomodationist policies.

After the war, American planning discourse moved either towards the European model of the modernist city or towards the planned new town, which the Americans had had a strong hand in developing. In the United States however the importance of market forces and the high rate of automobile ownership meant that long-standing practices of market-led suburbanization based on single family houses soon reasserted itself as the predominant force while urban modernism began to wane.[9] As much as the British, one of the main reasons why the American planners wanted an alternative to the high-rise city was an understanding that, given a choice, the public would never accept living in mega-structures; these were built by public housing authorities in the US for those unable to access the private market. The growing interest among planners in the new town by the late 1950s was supported by a standing proto-environmental critique of uncontrolled suburbanization under the banner of regional planning, dating back to the 1920s when an iconoclastic group of planners, the Regional Planning Association of America (RPAA) which included the critic Lewis Mumford, had foreseen the despoliation of agricultural and scenic areas by uncontrolled suburbanization (Luccarelli 1995). The RPAA also built prototypes, most famously Radburn, New Jersey, dubbed by its planner-architect Clarence Stein as 'a town for the motor age'. It opened in 1927 and was followed by a small number of greenbelt towns built by the federal government's Resettlement Administration during the Great Depression of the 1930s.[10] The RPAA suggested broad principles of landscape design, both aesthetic and ecological in their concept of regional planning.

---

9 It must be pointed out, however, that without infrastructural investments by the federal government, suburbanization could not have proceeded at anywhere near the pace or magnitude that transformed the country into a suburban nation (Altshuler 1983).

10 These were seen as experiments and mass suburbanization after the war was privately financed and market-driven, though there was a return to new town ideas in the 1960s when two famous best-practice examples, Reston, Va (1963) and Columbia, Md (1964) were built.

Figure 4.11 An urban motorway around the centre of Stevenage New Town. Note the railing in the central reservation. Pedestrians must cross using the subway (tunnel)

Source: Photograph by Mark Luccarelli, 2010

Figure 4.12 Despite the mid-rise, the centre of Stevenage New town has the sterile look and feel of a suburban shopping centre

Source: Photographs by Mark Luccarelli, 2010.

Exemplified in the work of Benton MacKaye, landscape principles were developed as a 'geotechnics' for limiting and shaping suburban and exurban development in line with natural features and geographical forces. Landscape became a force to be harnessed by the planner in bringing life and form to a new kind of regional multi-nodal city (Luccarelli 1995, Minteer 2006), but it remained a vision awaiting realization.

The British, on the other hand, were both practical and systematic in their drive to employ the new town. Sir Patrick Abercrombie drew up his London Plan in 1944 in response to the expected growth of the London region at the end of the war. In 1946, the New Towns Act was passed. It led to the establishment of a greenbelt around Greater London and the designation of eight new towns just beyond the belt between 1946 and 1955. The plans envisaged 50,000 inhabitants per new town, though the total design objective of 400,000 was substantially exceeded and accounted for a sizable percentage of suburban growth (Hall 1999).

The British new towns incorporated some garden city ideas, especially the prevalence of parks, terraced (row) housing and private gardens. But they were also sharply influenced by modernist design principles, particularly around the incorporation of motorways (freeways) into the landscape and in the development of highly formal and overly planned town centres (Figures 4.11 and 4.12).

One visit to the English garden cities with their picturesque parks, romantic cottage architecture and traditional town centres will tell you that they have a very different sensibility than that of the modernist/functionalist new towns. The perception of similarity and historical continuity remains because the primary issue, for a generation now, has been suburbanization itself. In the broadest sense as a model, the garden city, especially when transformed into the garden suburb as in Hampstead, did little to alter the reigning suburban ideology. In fact, the British new town centres did not share much of the garden city idea, much less the American regional planning tradition. New town centres have more the feel of a shopping centre surrounded by parking lots than that of traditional town centre, which had characterized the early garden cities (Figure 4.13).

Parks in the garden cities were of two sorts; picturesque parks on the periphery or in residential areas provided pastoral retreats (Figure 4.9) or they were designed to augment public space in a traditional urban centre, reinforcing the urbanist tendency of the garden city (Figure 4.13). On the other hand, new town parks contain many virtues but they make the most visual sense when seen as part of a unified landscape of the highway, the high-rise and the shopping centre. They provide a natural backdrop to these defining features of the highway-city. In fact, the parks can be quite pleasant, but they are largely places one passes through on the way to the shopping centre (Figure 4.14). The highway and the shopping centre are to the back of these photographs.

The problems with British exurban development are magnified 100 times in the USA where the new town idea of coordinated highway-centred development became widely adopted by private developers. Housing complexes were built in

Figure 4.13 Town centre of Letchworth garden city (top) and Welwyn Garden City (bottom)

Source: Photographs by Mark Luccarelli, 2010.

Figure 4.14 Park near the centre of Stevenage New Town

Source: Mark Luccarelli, 2010.

relation to shopping centres and office centre complexes. Suburbanization morphed into the 'edge city' phenomenon (Garreau 1992); a de-centred city encompassing bits and pieces of undeveloped landscape, 'campuses' and complexes of residence, commerce and industry, all made accessible by a network of limited access highways. The edge city had become a 'technoburbia', a centreless centre of industry, commerce and residence – a new kind of city based on hyper-mobility (Fishman 1988).

Postmodern Planning Discourse and the Burial of the Garden City

Even before urban critics and historians had reclassified metropolis as technoburbia, American planning discourse had overturned modernist planning tropes. The target of critique was the automobile hyper-mobility, which was tied to the loss of pedestrian-centred urban cores. The garden city, the RPAA's regionalism and the new town all seemed of a piece; the pre-war forerunners to a modernist planning discourse seen as the major protagonist in an assault against the city and against urbanism (Jacobs 1969). In the newly felt enthusiasm for the centre, landscape seemed to be in the way and the garden city was a seen as a misstep (Whyte 1968, 1990).

In Scandinavia, post-war planning became emblematic of other sorts of missteps. In Sweden, urban modernism prevailed as social democratic governments built huge suburban housing complexes after the war. The movement of the working class out of their wards in the centre city proceeded at a brisk pace. There was also a slower drift of the middle class to suburbs of semi-detached and detached private houses. The advantage, whether by design or accident, of the Scandinavian outcome is that unlike in the USA, the viability of the central city was not compromised by diffusion. The disadvantage is that the highly anonymous modernist housing complexes that were built have proven to be incubators of alienation and social unrest, outside of Stockholm as well as Paris, when the native working class moved out and Asian and North African immigrant populations moved in. The same housing ideal characterized pre- and post-war Oslo, at first under the auspices of the state-supported cooperative housing association (OBOS) and along lines of the Swedish model introduced at a famous exhibit in Stockholm in 1930 and envisioned in a master plan for greater Oslo in 1950 (Lorange and Myhre 1991: 142).

Interestingly, the garden city idea, first introduced at Ullevål Hageby in the 1920s, remained a part of the planning discourse in Norway, though not in the way originally intended. In 1948, at a conference in northern Norway at Tromsø, garden city principles were undoubtedly embedded in the context of comprehensive regional planning for both rural and urban Norway, along lines suggested by the RPAA planning model, Lewis Mumford's *The Culture of Cities* (1938) and as exercised by the Tennessee Valley Authority in the 1930s. Comprehensive regional planning by Mumfordian principles was deeply concerned with landscape and with the social aesthetics of residential areas. Aesthetics were understood to be important in developing a keener sense of place, and identification with place was understood to have important social and political implications. The Norwegian planners failed to garner state support and the government ended up advancing a new county plan based on area analysis undertaken by economists. Master plans for Oslo in the 1950s and 1960s emphasized modernist, high-rise new towns backed by an integrated suburban metro system built in the broad valley east of the city, with 50,000 units built there by 1970 (Lorange and Myhre 1991: 146–9).

At the same time, the suburban wish for a rural or garden setting proved to be stronger in Norway than in Sweden, which meant that there was less production of planned high-rise complexes. The opportunity to achieve this aspiration for privacy and a garden-setting grew as car ownership increased at the beginning of the 1960s. In addition, in common with many North American cities and unlike Stockholm and other European ones, flight from the city left parts of Oslo depopulated and run-down, until they filled up with immigrants in the 1970s on their way to gentrification in the 1990s.[11]

A more dispersed and relatively lower density suburbanization meant that rail traffic declined for a time – it is currently enjoying a renaissance – and road building increased from the 1960s through the 1980s, especially as oil money began to fuel metropolitan growth. Some garden city/new town housing principles were applied to small auto-dependent subdivisions that came to dominate suburban development after the 1960s. Not really new towns, they might be called 'new villages', though they are really nothing more than elaborated planned unit developments with some communal features. Like the new town, they took from the garden city the terraced (row) house, provision for common areas and the development of pedestrian footpaths. Like the new town, they took from the garden city the terraced (row) house, provision for common areas and the development of pedestrian footpaths. But they lacked the scale and density to support pedestrian activity or transit. In a way, these suburban settlements were a modern reworking of the Scandinavian landscape of settlement: a clearing in the forest remote from the world.

Conclusion

We have argued that the complexities of the garden city have been lost to planning discourses. In the period 1945–75, modernist-planning discourses absorbed aspects of garden city thinking into the new town ideas. On the one hand, this was understandable because the garden city had much in common with the new town and the 'new village' and other aspects of modernist planning; the value of suburban diffusion and an emphasis on accessing cheap land at the periphery of existing urban centres. Its emphasis on building *de novo* satellite towns seemed perfectly compatible with the new towns in Britain and the even lower density 1960s Norwegian housing 'regionalism' or auto-centred dispersed clustered housing. For all of these reasons it has been heavily criticized. Indeed, the garden city in this case is guilty by association with suburbia. Suburbs are often understood as outmoded forms (Alexander, Ishikawa, Silverstein 1977) and the garden city as a model based as Howard put it on 'the perfect marriage of countryside and city' has naturally been of little interest. The main significance of a garden city today seems to be as a negative example. In our view, then, 1970s postmodern planning badly conflated the garden city with suburbanization in general and therefore missed the former's emphases on landscape design and pedestrian orientation, qualities that make it compatible with urbanism. However, repositioning the place of the garden city in planning history is

11 Indeed, a slum-clearance project for the now gentrified old working class section of Grunnelökka was proposed in 1936 though never realized (Lorange and Myhre 1991: 142).

less important than recovering its aesthetic-social qualities. These, we believe, can make a useful contribution to green urban discourse.

Ullevål Hageby embodies certain lessons for landscape urbanism: define the area with strong borders; refer to the vernacular architecture/landscape; develop a certain amount of homogeneity in architecture through restoration and preservation of the existing housing stock and by building new areas with a certain unity of character; and include a central area or gathering point and develop a mixture of private, public and semi-public spaces, including parks of both pastoral and civic types.

Bibliography

Abercrombie, P. 1945. *Greater London Plan 1944*. London: HMSO.

Alexander, C., Ishikawa, S. and Silverstein, M. 1977. *A Pattern Language: Towns, Buildings, Construction*. New York: Oxford University Press.

Alpers, P. 1996. *What is Pastoral?* Chicago, IL: University of Chicago Press.

Altshuler, A. 1983. 'The intercity freeway', in *Introduction to Planning History in the United States*, edited by D.A. Krueckeberg. Chapel Hill, NC: Rutgers University Press.

Beatley, T. 2010. *Biophilic Cities: Integrating Nature into Urban Design and Planning*. Washington DC: Island Press.

Bourdieu, P. 1984. *Distinction: A Social Critique of the Judgment of Taste*. Cambridge, MA: Harvard University Press.

Braudel, F. 1949. *The Mediterranean and the Mediterranean World in the Age of Philip II*. New York: Harper & Row (Translated from the French by S. Reynolds and first published in English in 1972).

des Cars, L. 2000. *The Pre-Raphaelites: Romance and Realism*. London: Thames & Hudson.

Cosgrove, D. 1998. *Social Formation and Symbolic Landscape*. Madison, WI: Univ. of Wisconsin.

Duncan, J. and Duncan, N. 2004. *Landscape and Privilege*. New York: Routledge.

Edwards, R. 2011. Prior's Hall Barn, Widdington, Essex(photograph). Available at: http://www.geograph.org.uk/photo/148378 [accessed: 15 August 2011].

Emerson, R. 1892. The Young American in Nature; Addresses and Lecture, volume 1 in *The Complete Works of Ralph Waldo Emerson*. Boston, MA: Houghton, Mifflin and company.

Fishman, R. 1988. *Bourgeois Utopias*. New York: Basic Books.

Florida, R. 2002. *The Rise of the Creative Class*. New York: Basic Books.

Cosgrove, D.E. 1998. *Social Formation and Symbolic Landscape*. Madison, WI: University of Wisconsin Press.

Garreau, J. 1992. *Edge City: Life on the New Frontier*. New York: Anchor Books.

Gifford, T. 1999. *Pastoral*. London: Routledge.

Gilpin, W. 1782. *Observations on the River Wye: And several parts of South Wales, &c. relative chiefly to picturesque beauty; made in the summer of the year 1770*. London: Printed for R. Blamire. Sold by B. Law and R. Faulder.
Hall, P. 1999. *Cities in Civilization: Culture, Innovation, and Urban Order*. London: Phoenix.
Hall, P. 2002. *Cities of Tomorrow*. 3rd edition. Oxford: Blackwell.
Howard, E. 1902. *Garden Cities of Tomorrow*. 1965 edition with a preface, by F.J. Osborn and an introductory essay by Lewis Mumford. London: Faber & Faber.
Isaacs, R. 2000. The urban picturesque: An aesthetic experience of urban pedestrian places. *Journal of Urban Design*, 5(2), 145–80.
Jackson, J. 1999. 'In search of the protolandscape', in *Landscape in America*, edited by G.F. Thompson. Austin, TX: University of Texas Press, 43–50.
Jacobs, J. 1961. *The Death and Life of Great American Cities*. New York: Random House.
Jacobs, J. 1969. *The Economy of Cities*. New York: Random House.
Kjeldstadli, K. and Myhre, J. 1995. *Oslo – Spenningens By*. Oslo: Pax Forlag.
Le Corbusier. 1933. *La Ville Radieuse: Éléments d'une Doctrine d'Urbanisme pour l'Équipement de la Civilisation Machiniste*. Boulogne (Seine): Éditions de l'Architecture d'aujourd'hui.
Lorange, E. and Myhre, J. 1991. 'Urban planning in Norway', in *Planning and Urban Growth in the Nordic Countries*, edited by T. Hall. London: Chapman & Hall.
Luccarelli, M. 1995. *Lewis Mumford and the Ecological Region*. New York: Guilford Press.
Minteer, B. 2006. *The Landscape of Reform*. Cambridge, MA: MIT Press.
Morris, W. 1890. *News from No Where*. London: Longmans.
Mumford, L. 1926. R'egions – to live in', in *Survey Graphic*, 7, 91–3.
Mumford, L. 1938. *The Culture of Cities*. New York: Harcourt, Brace.
Mumford, L. 1963. *The Highway and the City*. New York: Harcourt Brace Jovanovich.
Norberg-Schultz, C. 1996. *Nightlands: Nordic Building*. Cambridge, MA: MIT Press.
Osborn, F. 1942. *New Towns after the War*. 2nd edition. London: J.M. Dent & Sons.
Osborn, F. 1969. *Green-Belt Cities*. 2nd edition. London: Evelyn, Adams & Mackay.
Osborn, F. and Whittick, A. 1977. *New Towns: Their Origin, Achievements and Progress*. 3rd edition. London: Leonard hill.
Ruskin, J. 1853. *The Stones of Venice*. 3rd edition 1908. London: Allen.
Shannon, K. 2006. 'From theory to resistance: Landscape urbanism in Europe', in *The Landscape Urbanism Reader*, edited by C. Waldheim. New York: Princeton Architectural Press, 141–61.
Sitte, C. 1889. *Die Städtebau nach der künstlerischen Grundsätzen*. Vienna.

Thorsen, K. 2011. *Byrom er ikke luksus*. Paper to the Bokvalitet i den moderne by Conference: Tab-Buk Conference, Oslo, 14 June 2011.

Treib, M. 1999. 'Nature recalled', in *Recovering Landscape: Essays in Contemporary Landscape Architecture*, edited by J. Corner. New York: Princeton Architectural Press, 29–44.

Unwin, R. 1909. *Town Planning in Practice.* London: Unwin.

Waldheim, C. (ed.) 2006. *The Landscape Urbanism Reader*. New York: Princeton Architectural Press.

Wall, D. 1999. 'Programming the urban surface', in *Recovering Landscape: Essays in Contemporary Landscape Architecture*, edited by J. Corner. New York: Princeton Architectural Press, 233–50.

Weller, R. 2006. 'An art of instrumentality: Thinking through landscape urbanism', in *The Landscape Urbanism Reader*, edited by C. Waldheim. New York: Princeton Architectural Press, 69–85.

Whyte, W. 1968. *The Last Landscape*. New York: Doubleday.

Whyte, W. 1990. *City: Rediscovering the Center*. New York: Doubleday.

Worster, D. 1985. *Nature's Economy: A History of Ecological Ideas*. Cambridge: Cambridge University Press.

PART II
Policies, discourses and dilemmas

Chapter 5
Can Cities be Green?

Peter Hemmersam

Urban sustainability is often reduced to a specific set of issues packaged in the notion of 'green city' or 'eco-city'. The architectural policy of Oslo, for example, emphasizes sustainability, and 'greenness' as marketable qualities which are exploitable in the city's self-promotion. But eco-cities are also experiments with future relationships between society, technology and nature. For architecture and urban design, the challenge is to look beyond immediate environmental 'fixes', and engage the public through representation, iconography and aesthetics and promote variety and experimentation with possible future green cities.

The notion of 'Green City', 'Eco-City' or 'Sustainable City' occurs in the policies of many cities, and has in some cases done so for a decade or two. Particularly since 2000, we have seen a tremendous rise in the awareness of the need for cities to focus on sustainability in general terms, and the struggle against climate change specifically, both in terms of limiting climate gas emissions, but also in terms of developing resilience in the face of a changing climate with potential negative consequences for cities in terms of flooding, increasing rainfall, drought and so on. An example of political initiatives towards sustainability is the C40 or Large Cities Climate Leadership Group that sprang out of the World Cities Leadership Climate Change Summit, organized by the Mayor of London in 2005, and seeks to build political alliances and promote exchange of knowledge and experience relating to the climate change challenge.[1] As competence is building and experiences are being made, cities approach the sustainability challenge along a number of parallel approaches, dealing with everything from policies on procurement of eco-technical services, building efficient and intelligent infrastructural systems, awareness building measures to CO_2 quota systems as well as addressing the appearance and design of the city itself.

An example of public policy promoting sustainability in architecture and urbanism can be found in the recent draft of a revised architectural policy for Oslo which was subject to a public hearing in 2009, and currently awaits further action and political approval. Contending that 'Oslo should grow attractive and sustainable', the city's official architectural policy lays out guidelines for the quality of public space, architecture, infrastructure and strategies for urban form

1 See http://www.c40cities.org/.

(PBE 2009).[2] The three key visions in the document all refer to sustainability in one way or another:

1. Through innovative everyday and monumental architecture, the capital is a national and international model for Norway's quality of life, democracy and *sustainability*.
2. The growth of Oslo has increased its attractiveness and *climate neutrality*. Its landscape, blue-green structure and urban areas with distinct urban design and architecture has been preserved and developed.[3]
3. The spaces, structures and buildings shared by the city and the national government have an especially high architectural and *environmental* quality.[4]

Concerning architecture, the most direct recommendations of the document deals with the importance of the reuse and adaptation of old buildings to meet current functional needs and energy requirements, with promoting a general awareness of zero or low-emission building in the population and with developing knowledge and expertise on energy efficiency within the construction industry. Regarding urbanism and urban design, it deals with the ecological performance of rivers, wetlands and other water bodies that flow into and through the city as well as adjacent green areas. Finally, it addresses the efficiency of infrastructure and public transport, and most significantly the question of the overall urban form, promoting the concentration of urban growth around public transport nodes.

While sustainability practices in architecture, at least to some extent, reflect a developing consensus within the profession, the draft architectural policy document for Oslo acknowledges that there is still some way to go before a comparable consensus on the contents of a sustainable urbanism is in place, and it suggests therefore that tests and experiments are still necessary. This could take the form of an architecture and urbanism exhibition. One such project has been initiated with an open architecture and planning competition for Furuset in the Grorud Valley. At the same time the document repeatedly emphasizes the need for developing standards and norms for environmental performance for both new and refurbished buildings and for urban environments. An example mentioned in the document is the LEED system, which grades and certifies individual buildings based on their projected environmental performance (prior to construction).[5] The system operates from a standardized model of what a 'green building' is, and employs a checklist of environmental metrics, such as energy saving, CO_2 emission and

2 All translations of the document are done by the author.

3 Oslo is sandwiched between the fjord and the protected forest, which makes up 70 per cent of the area of the municipality, and a series of brooks and rivers connect these natural features.

4 Italics by the author.

5 LEED = Leadership in Energy & Environmental Design.

water efficiency. The LEED certification also promotes other standards to cover wide aspects of the design, conversion and construction of buildings.

A problem with the call for norms and standards becomes evident in the first chapter of the document, with the introduction of the concept of 'sustainable development'. This concept necessitates the adaptation of a long-term perspective, and that evaluation of environmental performance must be based on an assessment of how objectives and targets are met over time and under conditions of use (Butters 2004, cited in PBE 2009). In other words: norms and standards as they are currently being employed, for instance in the LEED system, are based on projections of performance in idealized conditions, not actual performance. These models do not reflect the variable that is the actual use of the building over time, and are unable to reflect whether users actually programme the technical energy saving systems properly. Also, their failure to consider that flexibility of use, which is generally considered a sound sustainability strategy, means that future tenants potentially would add additional technical systems (like air conditioning) (Farmer and Guy 2005). Adding to the complexity of projecting and evaluating performance, which is fundamental for norms and standards, the document continues: 'sustainability is a dynamic concept which is continuously undergoing revisions according to changes in society, new technical solutions and public demands' (PBE 2009: 7). While standards may make perfect sense in today's society, there is no guarantee that future generations will see things the same way, which means that building with a lifespan of 40 to 100 years, may have to respond to a number of entirely different sustainability approaches and regimes.

The sustainability agenda that the document subscribes to is rooted in a very specific discourse. The words of the title of the draft architectural policy, 'growth' and 'sustainability', places the document squarely in the dominant trend in environmental thinking within politics, as exemplified by the UN Brundtland Commission report's call for 'sustainable development', which was defined as 'development that meets the needs of the present without compromising the ability of future generations to meet their own needs' (WCED 1987). The report signalled an important shift in environmental thinking, where the sustainability agenda was inserted into mainstream economic and politics, and a break from the proposals of the pioneering environmental movement for a more 'eco-centric' approach based on alternative, anti-capitalist values. It can be argued that this shift laid the foundation for a more technologically oriented and philosophically positivist 'techno-centric' approach to sustainability in architecture and urbanism. This approach tends to define problems and goals from a natural science perspective, using measurable quantities, eventually opening the way for the development of sustainability policies based on standards of performance. The literature of the techno-centric approach, which today has come to dominate the discourse in architecture, tends to be normative and prescriptive, calling for standards.[6] The

6 Such as the Passive House Standard, which specifies an ultra-low energy consumption for buildings.

underlying logic is that standards and systems for sustainability classification will ensure rapid and uniform achievements in the movement towards sustainable architecture and cities.[7] The problem with standards and classification systems is that they rely on a limited set of metrics and are biased towards the physical and quantifiable attributes of the built environment while overlooking questions of use, social equality, cultural preferences, and so on. Looking closer at classification systems, one finds that there is often little consensus about what criteria they should be based on, and there are often serious methodological problems, such as developing models of performance aspects that are susceptible to variations as a result of changes in use and programming (Farmer and Guy 2005).

As is the case with architecture, the questions of sustainability and climate neutrality in urbanism and the design of cities are often reduced to a discussion of a limited range of factors. Dealing with urban design, they include energy and water savings, the greening of roofs for water retention, urban agriculture and food production, the reduction of automobile use and the associated shift to public transport, cycling and walking, increased urban density and the fight against urban dispersal. For many of these issues, there is often a lack of scientific consensus about their actual efficiency in achieving sustainability and about their general applicability in different urban situations, where environmental and climatic conditions may vary and densities, cultural and consumption patterns may be hugely different. The introduction of the concept of sustainable development in the recent draft of the architectural policy for Oslo informs us that success can only be evaluated over time, and one would then encounter shifting social and cultural contexts that would inform future discourses on sustainable development.

Looking more closely at the sustainability agenda in architecture and urbanism, we find that there are not only one, but many different and often competing 'sustainability agendas', as Simon Guy and others have demonstrated (Guy and Farmer 2001), each with a slightly different emphasis on technology, community, health, and so on. Critically, they range from the technologically oriented agenda that dominate current thinking on building construction and urban design, to anti-capitalist and anti-technological ideologies, and to more socially- and community-oriented ways of thinking. These approaches are exemplified, on the one hand, by the low-tech, do-it-yourself approach, with a strong emphasis on community values, amateurism and anti-authoritarianism that is demonstrated in many smaller communities in several countries (such as the Findhorn Eco-village in Scotland or Christiania in Copenhagen, the New Age or spiritually oriented projects such as Arcosanti in Arizona or Auroville in India), to, on the other hand, the academically defined scientific projects, such as those promoting healthy living environments and the high-tech and space technology-inspired self-sustaining 'closed loop' projects, for instance, employing food production in the form of hydroponic or

7 For instance LEED or BREEAM (Building Research Establishment Environmental Assessment Method) which is being adopted as a standard by the construction industry in Norway as a tool for classifying and marketing the 'sustainability' of buildings.

aeroponic systems, where food is produced indoor in laboratory like conditions without the need for contact with soil.

Despite the multitude of agendas, the techno-centric discourse is almost entirely dominant within urban planning, and some very specific policies are being adopted by a large number of cities. These policies follow a parallel trajectory, exemplified by the most important strategy when it comes to physical urban planning: the Compact City.

The Compact City

The title ('The Sustainability and Growth of Oslo') as well as the contents of the latest draft architectural policy serve as illustrations of how the sustainability agenda has been adopted by mainstream urban planning policy. In the 46-page document, the word '*bærekraft*' ('sustainability') occurs 94 times, '*miljø*' ('environment') 80 times, '*klima*' ('climate') 46 times and '*energi*' ('energy') 24 times.

In terms of urban design, the main approach that the document promotes is the adoption of a 'Compact City' strategy, promoting dense, mixed use and energy-conserving forms of urban design around public transport nodes or arteries. This approach represents an attempt to undo half a century of automobile dependent planning, which has resulted in a widely dispersed urban region, and a reliance on individual car use resulting in congestion and substantial areas of poor environmental qualities adjacent to roadways. Emissions from vehicles also make up a large part of the total carbon emission of the city. The Compact City is a long-term strategy for changing the form of the city. It is based on the observation of a correlation between the environmental impact of cities (in terms of climate gas emission) and urban form. For instance, the residents in New York City, who to a large extent rely on public infrastructure and live and work in small spaces, have a carbon footprint which is only about a third of the national average in the United States (Kerr 2010). The concept has been particularly influential in an American context, where urban dispersal and the development of widespread metropolitan territories with little or no relation to central city cores has been a cause of worry for planners since before the Second World War. The development of this approach has been associated with the concept of Smart Growth and since the early 1970s with New Urbanism, a movement which saw the creation of community as a prerequisite for 'repairing' the fragmented contemporary urban landscape, and uses traditional and pre-modern forms of urban design to dissuade people from using cars in their daily life.[8] A newer variant of the concept is 'Transport Oriented Development', which

8 'Smart Growth Principles: 1. Create Range of Housing Opportunities and Choices, 2. Create Walkable Neighborhoods, 3. Encourage Community and Stakeholder Collaboration, 4. Foster Distinctive, Attractive Communities with a Strong Sense of Place, 5. Make Development Decisions Predictable, Fair and Cost Effective, 6. Mix Land Uses, 7. Preserve Open Space, Farmland, Natural Beauty and Critical Environmental Areas,

promotes the use of public transport by developing dense neighbourhoods around public transport stops, following the pattern of post-war new town development as exemplified by Stockholm, Copenhagen or Oslo. This set of strategies, here loosely bundled together under the heading 'Compact City', has over two decades become a primary instrument in adjusting planning policies and redirecting urban growth in many cities in line with addressing the general environmental challenge and responding to pressure to reduce climate gas emission. The Compact City strategy is however a case of policy racing ahead of factual, study-based research (Breheny 1996). For the last decade or more we have seen academia trying to make up for this deficiency – often with ambiguous results that have prompted calls for additional research (Dempsey and Jenks 2010). There are for instance no practical means of investigating whether the promised environmental benefit of the Compact City (reduction in energy consumption and CO_2 emission) is achievable. Other questions that remain unanswered are: what is compact? Looking closer at the issue, the answer varies wildly across different cities, from Australian suburbia to South and East Asian Megacities. Do people really want to live in compact cities, or do they prefer a suburban way of life that seems to be associated with increased affluence? What does the compact city really mean in terms of social sustainability and equity – is there a potential of aggregating social disparity as a result of spatial policies based on the sustainability imperative? Does mixed use really encourage alternative means of transport? What is the sustainability outcome when the Compact City strategy is applied in different contexts (Dempsey and Jenks 2010)? This uncertainty, coupled with the lack of a consensus on a definition of a Compact City may, however, be seen as a kind of dubious advantage, a flexibility that in a certain sense enables the Compact City sustainability approach to be easily associated with already existing policies and problems, making it much easier to adopt. Thus, Compact City policies may appear as dynamic and forward looking, while they may in fact be little more than wishful thinking, or simply traditional real estate development strategies in new guise; or, in some cases, they may just repeat already established, modernist new town urban planning principles that have been in place for decades.

In the case of Oslo, one can ask specific questions about what the adoption of a Compact City strategy actually means. The post-war development of Oslo to a large extent took the form of new towns developed around metro stations, and thus it already matches some of the principles of the Compact City and Transport Oriented Development. Therefore, the adoption of a Compact City strategy in Oslo merely strengthens the existing urban form, and has little effect on outlying areas beyond the city borders where car-dependent suburban expansion has occurred in the last couple of decades. A question also remains as to whether Compact City style developments of public transport hubs will appear as attractive living environments, or if potential residents will simply migrate out over the city border

---

8. Provide a Variety of Transportation Choices, 9. Strengthen and Direct Development Towards Existing Communities, 10. Take Advantage of Compact Building Design' (SGN 2010).

to the neighbouring municipalities that offer the lower residential densities that appear to continue to be attractive to large segments of society.

In the complexity of issues associated with the notion of Compact City, from social to infrastructural and economic, cultural and environmental, it is even conceivable that sustainable achievement in one field may actually negatively affect development in another area, suggesting that Compact City policies are actually made up of potentially conflicting interests. While the cost of infrastructure provision and car ownership ratio may be reduced in a Compact City situation, compact living could conceivably lead to overcrowding and the negativity associated with the loss of recreational space; density may be inversely correlated with identification and a sense of community belonging amongst inhabitants of a specific neighbourhood. Furthermore, if one accepted living with these negative aspects of densification, the contribution to the reduction in greenhouse gas emissions and urban sustainability may turn out to be insignificant: recent research indicates that residential energy consumption seems to be more a question of lifestyle than the actual density of an urban environment (Jones et al. 2010).

Eco-cities

Many cities are now promoted as 'green' or 'eco' or 'sustainable'. Primarily, this is associated with new urban developments, in proximity to existing urban areas, or as entirely new cities developed on virgin land. As there is a lack of a shared definition, the claim can neither be proved nor disproved. Rather, we are talking about degrees of sustainability (Jones et al. 2010). In a sense, this makes the sustainability agenda valuable and useful to cities, as they can potentially use it to rebrand other existing policies and issues (Guy and Farmer 2001). Old policies can be redressed in sustainability rhetoric, acquiring new symbolic value, but in the process concepts like 'eco-city' or 'green city' become dangerously diluted or devoid of meaning. In short, language loses its meaning and becomes a prop for established practices. A related problem is apparent when we look at what actually comprises an eco-city. An eco-city is a city designed with the intention of reducing its environmental impact, minimizing pollution and reducing consumption of energy and finite resources. The term 'eco-city'[9] in its current use also involves the limitation or elimination of climate gas emissions. It not only refers to cities that do not consume resources and limit damages, but also to situations where they actually generate energy beyond their own needs, actively clean up polluted air, soil and water of the local ecological region or consume and process more climate gas than they emit.

When looking at existing eco-cities, one finds that the most widely disseminated and influential 'best practice' examples are either small residential developments built *de novo* or projects located on sites in the urban periphery. There are several

9 The term was originally introduced by Richard Register (1987).

reasons for this: one important factor is that the development of entire towns and cities is a long and complex undertaking, and from the time the sustainability imperative made its way into mainstream politics in many countries around ten years ago, to the day eco-cities are a fact on the ground, a considerable time will have passed. A difficulty in the development of eco-cities is the complexity of interdependent issues that make up sustainable development, from energy production, food production, economic viability and collective cultural issues, as well as personal consumption habits. Some of these may be addressed at a local or city scale, while others require a regional, national or even global approach. In an established urban situation, close to or within existing cities, a large number of property owners and other stakeholders would have to be included in the process. This complexity could potentially hinder consensus building that would be necessary for such an endeavour. The development of eco-cities is still an uncertain science, based on assumptions and projected performance, and any such project requires a degree of risk taking and experimentation, and any realized project would essentially be an experiment. This degree of uncertainty distinguishes eco-cities from traditional real estate development logics, and may hinder the adaptation of environmental technologies and principles on a comprehensive scale in existing urban neighbourhoods and city districts with large number of stakeholders and potentially conflicting interests.

Some of the most celebrated eco-cities are not yet realized: for instance the Dongtan eco-city in China planned by the international consulting engineering company Arup, which is also involved in a series of other eco-city projects in China. The city was to be developed on agricultural land on the large island of Chongming in the Yangtze River Delta and connected to Shanghai around 40 kilometres away. The city was planned for a population of up to 500,000 by 2050. It was to be self-sufficient with energy and water and have a small ecological footprint. Only electrical and hydrogen powered cars were to be allowed in the city centre and the urban form of Dongtan was to be developed along the lines of Transport Oriented Development with all residences no more than seven minutes from public transport stops. Little or no waste would be produced and wind turbines, solar power and biomass processing would provide energy for the city's low-energy buildings. Intensive industrial food production would mean that the city could be built with no drop in the yield of the existing agricultural area. The first phase of the city was planned and promoted as a major feature of the World Expo in Shanghai in 2010, but it was put on hold or cancelled, and was perhaps little more than a marketing stunt. If it had been realized, Dongtan would have been a 'relatively low-density, rich, upper-middle-class suburb for commuters to Shanghai – anything but socially and economically sustainable' (Jenks 2009: 66). A similar claim could be made about the celebrated Vauban residential area in Freiburg, Germany, which is a realized and functioning urban district on the outskirts of the city with a distinct sustainability profile. All houses are low or 'ultra-low' energy buildings; power is produced by burning wood chip and with the help of solar voltaic panels. Vehicular access is limited in the district, and car ownership is very low compared to the German average. The social mix of the area is however not representative of the overall population. There are some spaces for students and an amount of subsidized housing however, mainly, the area houses families with higher education and children.

> While pretty much every city can point to one thing that is really good in terms of its sustainability or energy, a few can point to two, some can point to three or four ... But virtually nowhere can really claim that they have set the target that is in line with the science [...] on climate change, [... nor can] they [claim to] have actually got a program that is going to deliver it within the time scale they have set, or indeed get anywhere near it. (Watts 2010)

Some of the most celebrated 'best practice' eco-cities around the world seem to narrowly showcase a few selected issues that they are really good at. Cycling in Copenhagen is being celebrated as a best practice case, but it is really an old tradition that is now being celebrated as a visionary strategy for other cities to follow: to 'Copenhagnize' cities is to promote facilities for pedestrians and cyclists in urban and transport planning.[10] The celebrated Bus Rapid Transit system in Curitiba is an integrated system of buses partly running on separate bus lanes, which stops at purpose designed stops and has an integrated pricing system. It has had spectacular success in reducing congestion and improving quality of life for the inhabitants of Curitiba, and has been copied across the world as a low-cost alternative to more expensive light rail or underground metro systems. In Curitiba, the system has been supplemented with other initiatives and land use guidelines, supporting urban intensification along bus corridors. However successful the project has been, it still only deals with a limited number of the entire range of issues that must be addressed in the pursuit of the sustainable city.

Masdar City, designed by the architectural practice Foster & Partners, in Abu Dhabi is the most publicized large-scale eco-city currently under construction. It is a six square kilometre area next to the airport in Abu Dhabi, and is planned to house a university, research facilities and up to 50,000 residents and 60,000 workers. It will be serviced by a new public transport system (the first in Abu Dhabi) and a revolutionary system of individual automated transport pods that will replace car traffic inside the development. It is designed to provide maximum protection from the hot climate, using low-tech solutions for shading and air ventilation. Energy will be provided by wind farms, architectural photovoltaic panels and a solar power plant, which will also power the desalination plant that will produce most of the development's drinking water. The development of Masdar City is apparently not affected by the recent financial crises, where developers would be expected to be wary of the extra cost of eco-technologies. This is no problem for Masdar which is backed by Abu Dhabi oil revenue, and sponsored by the government in a less than democratic nation. As Peter Hall points

---

10 See www.copenhagenize.com.

out, Masdar '... is probably one of the few places where you can hope to achieve this' (BBC 2010). The claim to sustainability for such a project is questionable; it may constitute a zero emission entity in itself, but it relies heavily on air travel and car infrastructure in its incredibly resource-consuming context and it remains to be seen whether it will be able to attract residents who would necessarily have to adapt to certain lifestyle changes. There is also the question of what the socio-economic profile of its residents will look like.

The above cases are examples of the 'best practice' model for propagating sustainability in urban design. In some cases, best practice models may function as blueprints for decision makers and designers, potentially resulting in questionable outcomes, where projects 'tick all the right boxes' but they avoid addressing context-specific problems and tapping into available local resources and potentials. In truth, this propensity of avoiding place- and culture-specific contexts is precisely the problem with the adaptation of 'compact city' strategies in many places around the world. This is clearly the case when we examine the circumstances of cities across the north-south divide. Compact City strategies are relevant largely to Western cities, and do not make much sense in the mega-city context of the developing world, where densities and levels of programmatic mixture already far exceed what we find elsewhere (Jencks 2009).

While the environmental and sustainability merits of the eco-city are at best fragmented or simply yet unproven, they may nonetheless play an important role as they function as practical laboratories in which technologies are tested in optimal surroundings at the same time as the application of technologies and designs are carried out in existing urban fabrics (BBC 2010). Best practise model eco-cities can also have an enormous direct impact on perceptions and policies. A prominent example is the cancelled Dongtan project by Arup, which although unrealized, inspired the British government to instigate the construction of ten new eco-towns across the UK to function as laboratories and to promote sustainable urban development. The eco-towns should have a minimum of 5,000 homes each, and combine environmental sustainability with a proportion of affordable housing. They should be low on carbon emission by promoting walking and cycling and locating homes in proximity of public transport. A number of projects have been initiated, and a second wave is being planned (DCLG 2010). The policy is not without critics, and while some of the eco-towns were planned as supplements to existing towns, others take the form of greenfield developments, inspired by the Dongtan project. The urban expansion into open agricultural land, disconnected from existing urban infrastructure and services, has raised serious questions as to the total sustainability of the eco-towns (Revill and Davies 2008, cited in Jenks and Jones 2010).

Attractive Cities

'A new dimension of cities' competitive positioning is their ability to internalize and control the resources with which they are endowed and which subsequently supply, consumption, and production' (Hodson and Marvin 2010: 214). The draft architectural policy can be seen as an element in the positioning of Oslo as an environmentally progressive city, making it more attractive and competitive by enticing tourists and businesses, as well as new inhabitants: in short, 'greenness' is a marketable quality, and an exploitable image designed to attract the 'creative' knowledge workers that cities think they need to compete in the New Economy. In an age of globalization and increasing inter-city competition, cities have for some time focused on producing attractive urban environments and pleasant inner city districts. They may take the form of redeveloped former industrial harbour fronts like the Aker Brygge in Oslo or the upgrading of central urban spaces such as the renovation of Karl Johans Gate in Oslo. The goal is aesthetic but also economic, attracting businesses, tourists and 'creative' inhabitants; these strategies enact the Compact City focus on the perceived quality of urban space. Compact urban environments have traditionally been associated with slum and social problems related to overcrowding, worn out facilities and lack of recreational space. While this may not be the case today in cities like Oslo, strategies for compaction and intensification of activities is often accompanied by an intensified and critical attention to liveability: the quality of urban space, the design and layout of buildings and open space in order to ensure views, lighting conditions and access to green areas.

The draft architectural policy argues for densification and this strategy may actually go hand in hand with existing strategies for new attractive and competitive urban development like the fjord city project, which is the urbanization scheme for the entire central waterfront in Oslo. It comprises up to 10,000 housing units and large cultural institutions like the new opera and Munch Museum. Both of these large cultural institutions are located in Bjørvika, which has been planned to become the new Central Business District of Oslo. Bjørvika, identifiable through its iconic 'Barcode' office towers, represents a planning strategy that seeks to make Oslo attractive to business by developing a 'modern' business district in close proximity to the central railway station which serves as the terminus of the airport express train. The area is being developed as a high-density complex of office, retail and residential buildings, replete with buildings up to 22 stories high and a total floor area approaching 1 million square metres, planned to accommodate up to 20,000 new jobs. The project is being promoted internationally by the City of Oslo and represents a case where planning policies promoting economic development go hand in hand with the Compact City model promoting sustainability.[11] The urban planning policy of the Blair government (the 'New Urban Renaissance') is a like example and the connection between the two projects has been more or less explicitly made. In this policy, urban densification and development of post-industrial areas are linked to sustainability and quality of urban environments and questions of the overall attractiveness and competitiveness of cities (UTF 1999).

11 The exhibition 'Oslo – The Fjord City Expo' has travelled to China, Germany, France, Italy and Russia.

Promoting Sustainability

The rhetoric of the draft of the architectural policy for Oslo is very different from the existing architectural policy from 2005 (PBE), which focused on aesthetic qualities of the city. This radical shift echoes the weight sustainability and climate neutrality has gained in policy circles and in the general population. Promotion of sustainability to the public is specifically mentioned in the draft architectural policy: 'the visitors [of a proposed architecture and urbanism exhibition] should become aware of the possibilities for building and buying climate-friendly houses' (PBE 2009: 45). Winners of the competition for the architecture and urbanism exhibition should be promoted in the same way as it has happened in other recent open architectural competitions: Norwegian Wood, which its host city, Stavanger, has used to promote itself as an innovative and environmental city and The National Tourist Routes project by The Norwegian Public Roads Administration, which consists of installations and road side tourist infrastructure designed by new architectural firms, and has developed into a kind of travelling exhibition of a younger generation of architects expressing the state of Norwegian architecture.[12] It has been used in promoting Norway and Norwegian architecture on an international arena (NPRA 2009).[13] In this way, part of the environmental emphasis in the document, and the call for experimentation can also be seen as an example of the adaptation of sustainability policies for the purpose of promoting the city as innovative, future oriented and 'green'. A conscious effort to develop this picture of Oslo associates itself with attempts to promote the city on the international stage and to attract talent, business and investment.

The open ideas competition for Furuset in Oslo is an effort to create a best practice model for 'climate efficient' urban development. This project forms part of a wider drive towards sustainable urbanism in Norway, which also includes the recently completed urban design competition for Strømsø in Drammen and the ongoing competition for Brøset in Trondheim. The purpose of these projects is to develop models for low-emission or zero-emission urban design, a concept that echoes the idea of Passive House (ultra-low emission buildings) or Zero Emission Building (that consume no energy at all, but potentially generate energy). These standards are usually associated with very well-insulated individual buildings with high-tech components controlling heat transfer between the exterior and interior of the building. The technology to achieve this goal on the scale of an individual building is available and relatively well established at this time, but any urban design aspiring to be carbon neutral or sustainable in a broader sense would have to deal with a much wider range of social, cultural, economic and political issues, ranging far beyond the physical limitation of the design of a neighbourhood. So, while ultra-low or zero emission might be a starting point for developing sustainable urban design, a credible design would have to reflect the fact that in many cases the relationship between design and sustainability is far from being causal due to

12 See http://www.arkitektur.no/?nid=5920.

13 See http://www.turistveg.no.

the complexity of issues involved, including collective and individual actions and decisions. The urban design developed at Furuset and other best practice cases may function as experiments with the interaction between environmental, economic and social sustainability, but their most important contribution may be to function as icons of, and visual reference to, a particular political agenda associated with sustainable development.

Shades of Green

In architecture, a distinction between 'green' and 'sustainable' projects has been proposed to tell between different approaches to sustainability apart (Cole 1999). In such a classification, 'green' would signify incremental improvements of the existing standards, while 'sustainable' is reserved for buildings that differ radically from conventions and establish entirely new standards of performance. Other versions of this differentiation include shades of green from 'light' to 'deep' to indicate the performance of a buildings on a scale of adaption to 'absolute' sustainability. This approach depends on some sort of measurable scale and consensus on which parameters are pertinent, and how projections and measurements of performance are carried out. This would require a universal adaptation of standards and Environmental Assessment Methods, like LEED and BREAM.

However, we find ourselves in a situation where models and best practice examples for cities are still not fully developed, and unlike architecture where a sort of consensus seems to be closer at hand, there is still no consensus on exactly what constitutes sustainability within urbanism and urban design. The call for standards, as we find it in the draft architectural policy, is intended to spark a drive within the construction business and planning and design professions to develop cost-efficient technologies to meet new demands and requirements. At the same time, the very same introduction of standards could hinder experimentation that still is absolutely critical in the development of cities that are not just 'business as usual', made sustainable in one or a few selected areas. A process of standardization and norms could also limit the possibility and motivation to adapt to local contexts and resources, as comparative (ranking) methods would have great difficulty in evaluating projects that put emphasis on alternative hierarchies of issues and a variety of sustainability agendas and leave little room for anything other than the prevailing techno-centric approach to sustainable architecture.

In delivering sustainability, an architectural policy is not enough. The problem is that there is little causality in the relation between urban form and sustainability, as 'Building sustainable developments does not mean that the residents have sustainable lifestyles' (Jones et al. 2010: 252). While achievements towards sustainability concerning certain factors may be facilitated and helped along by architecture and urban form, addressing a wide range of social, economic and cultural issues, concerning everyday actions, attitudes and lifestyles choices of citizens are of absolute importance (Brand 2004). Addressing these issues would require a multitude of approaches, many of which lie beyond architecture and urban policy. There is nonetheless an important role for architecture and urban design in delivering 'images' that visually demonstrate and give reference to a different relationship between people, the built environment and architecture. The sustainability agenda in urbanism in the form of the green city or eco-city can potentially contribute to the formation of a new consensus on economic, social and urban development: 'The sustainable city can be understood as an ideal – a particular fixation of social aspirations and spatial meaning, with the power to reorganize urban purpose and legitimate public action in the name of the common good' (Brand 2000). This common good refers to the idea of a neutral notion of an objectively sustainable city, which is reflected in the current discourse on sustainable urban development. As we have seen, such an agenda is susceptible to degrees of manipulation, and the sustainable planning and architecture agenda plays out very differently in various cities, according to how it is being absorbed into existing policies and interests. It is important to look 'behind the interests promoting a narrower definition of what we'll call eco-urbanism as a specific temporal and spatial response to the challenges of climate change and resource limits' (Hodson and Marvin 2010: 208). While a techno-centric positivist approach dealing with targets and climate neutrality (measured in tons) currently has the upper hand in the current political and planning milieu, other approaches and 'logics' in understanding sustainability in urbanism are still debated, and perhaps the controversy itself is both a critical aspect of the urban situation and public space itself, and a thing that is both inevitable and crucial in the relationship between the city and the topic of sustainability (Mostafavi 2010).

An architectural policy may be a surprising place to find such an emphasis on sustainability. The policy proposes 'best practice' projects and competitions, supporting experimentation in the form of urban design projects, and the 'emphasis on projects rather than policies is recognition of the value of projective possibilities for the physical development of the region. This type of speculative design is a necessary precondition for making radical policies that are embedded in imaginative and anticipatory forms of spatial practice' (Mostafavi 2010: 47). 'Green city' is an open concept, that should not be technically defined (and perhaps not be defined at all), but rather be open to varying contexts, important agendas and other changes in society; consequently, architecture and urban design should experiment and 'develop the aesthetic means – the projects – that propose alternative, inspiring, and ductile sensibilities for our ethico-political interactions with the environment' (Mostafavi 2010: 50). The draft architectural policy for Oslo mentions the greening of buildings with vegetation as one of several strategies towards sustainability, even though to the visitor Oslo appears to be an exceptionally green and low-density city. Greening is a widespread technology within the currently predominant approach to sustainability in architecture and urbanism, but setting aside the fact that the environmental credentials of vegetated buildings may, in some cases, be questionable (Zardini 2010), the strategy plays an important part in visually demonstrating that individual buildings can address environmental issues. Greening of buildings represents a rhetorical device, visualizing potential changing relationships between the built form of the city, society and the natural environment. By introducing visual icons of change into the city, new meaning is introduced in the everyday urban setting.

The concept of sustainability lacks definition, and there is no such thing as a truly sustainable city. Frankly, cities can only become sustainable to a degree (Cooper et al. 2009), but it is possible to develop strategies that can bring about sustainability in an incremental way. Cities can be 'green' in promoting aspects of the sustainability agenda in external city branding, promoting economic development, but also in relating to their citizens by opening up for possible futures. Hereby, the notion of the green city or eco-city may acquire importance beyond strictly environmental improvements by providing a rhetoric and building acceptance for radical changes and an atmosphere where very different experiments with new possible relationships between technology and nature can take place in the city.

Bibliography

BBC 2010. *Eco-City Limits* [Online: BBC Radio 4, Costing the Earth]. Available at: www.bbc.co.uk/iplayer/console/b00rm3pr [accessed: 14 November 2011].

Brand, P. 2000. 'The Sustainable city as metaphor: Urban environmentalism in Medellin, Colombia', in *Compact Cities: Sustainable Urban Forms for Developing Countries*, edited by M. Jenks and R. Burgess. London: Spon Press.

Brand, P. 2004. *Urban Environmentalism in Context: Case Studies from Birmingham (UK), Lodz (Poland) and Medellín (Colomb*ia). City Futures Conference: University of Illinois at Chicago, 8–10 July 2004. Available at: http://www.uic.edu/cuppa/cityfutures/papers/webpapers/cityfuturespapers/session8_6/8_6urbanenvironmentalism.pdf [accessed: 12 April 2010].

Breheny, M. 1996. 'Centrists, decentrists and compromisers: Views on the future of urban form', in *The Compact City: A Sustainable Urban Form?*, edited by M. Jenks, K. Williams and E. Burton. London: E & FN Spon.

Butters, C. 2004. *Et helhetlig verktøy for evaluering av bærekraft*. Plan, 1, 4–11.

Cole, R.J. 1999. Building Environmental Assessment Methods: Clarifying intentions. *Building Research and Information*, 27(4/5), 230–46.

Cooper, R., Evans, G. and Boyko, C. 2009. *Designing Sustainable Cities*. Oxford: Wiley-Blackwell.

Dempsey, N. and Jenks, M. 2010. The future of the compact city. *Built Environment*, 36, 116–21.

DCLG 2010. Eco-towns [Online: Department for Communities and Local Government]. Available at: http://webarchive.nationalarchives.gov.uk/+/http://www.communities.gov.uk/housing/housingsupply/ecotowns/ [accessed: 14 November 2010].

Farmer, G. and Guy, S. 2005. 'Hybrid environments: The spaces of sustainable design', in *Sustainable Architectures: Cultures and Natures in Europe and North America*, edited by S. Guy and S.A. Moore. New York: Spon Press.

Guy, S. and Farmer, G. 2001. Reinterpreting sustainable architecture: The place of technology. *Journal of Architectural Education*, 54(3), 140–8.

Hodson, M. and Marvin, S. 2010. Transcendent eco-cities or urban ecological security?, in *Ecological Urbanism*, edited by M. Mostafavi and G. Doherty. Cambridge, Massachusetts, MA: Harvard University and Lars Muller Publishers, 208–15.

Jenks, M. 2009. 'From the compact city to the defragmented city', in *Eco-Urbanity: Towards Well-Mannered Built Environments*, edited by D. Radovic. London and New York: Routledge, 63–74.

Jenks, M. and Jones, C. 2010. 'Issues and concepts', in *Dimensions Of The Sustainable City*, edited by M. Jenks and C. Jones. Dordrecht, Springer, 1–20.

Jones, C., Jenks, M. and Bramley, G. 2010. 'Complementarities and contradictions', in *Dimensions Of The Sustainable City*, edited by M. Jenks and C. Jones. Dordrecht: Springer, 239–56.

Kerr, L. 2010. 'Shrinking Gotham's Footprint', in *Ecological Urbanism*, edited by M. Mostafavi and G. Doherty. Cambridge, MA: Harvard University and Lars Muller Publishers, 554–9.

Mostafavi, M. 2010. 'Why ecological urbanism? Why now?', in *Ecological Urbanism*, edited by M. Mostafavi and G. Doherty. Cambridge, MA: Harvard University and Lars Muller Publishers, 12–55.

PBE 2005. *Vakker By – Handlingsprogram For Estetikk og God Byarkitektur*. Oslo: Plan – og bygningsetaten [Agency for Planning and Building Services].

PBE 2009. *Oslos Bærekraft Og Vekst – Overordnet Arkitekturpolitikk For Byen Og Hovedstaden*. Oslo: Plan – og bygningsetaten [Agency for Planning and Building Services].

Register, R. 1987. *Eco-city Berkeley: Building Cities For a Healthy Future*. Berkeley, CA: North Atlantic Books.

Revill, J. and Davies, C. 2008. 'Secret' eco-town plans spark protest. *The Observer* [Online, 10 February]. Available at: http://www.guardian.co.uk/society/2008/feb/10/communities.planning1 [accessed: 25 August 2011].

SGN 2010. *About Smart Growth* [Online: Smart Growth Network at The National Center for Appropriate Technology]. Available at: www.smartgrowth.org/about/default.asp [accessed: 13 November 2010].

NPRA 2009. *Nasjonale turistveger i New York* [Online: The Norwegian Public Roads Administration]. Available at: http://www.vegvesen.no/Om+Statens+vegvesen/Media/Nyhetsarkiv/Nasjonalt/117150.cms [accessed: 16 April 2009].

UTF 1999. *Towards an Urban Renaissance*. London: Urban Task Force and Routledge.

Watts, M. 2010. *Klimatilpasset byutvikling* [Online]. Presentation to the conference, Framtidens Byer 2010, Drammen, 2–3 November (Oslo: Ministry of Environment). Available at: www.youtube.com/user/Framtidensbyer#p/u/62/YkFNyhWjGpI [accessed: 25 August 2011].

WCED 1987. *Our Common Future: Report of the World Commission on Environment and Development*. Oxford and New York: Oxford University Press.

Zardini, M. 2010. Different ways of becoming green, environmentally friendly and sustainable. *Lotus*, 140, 113–21.

Chapter 6

Shifts in Planning Ideology and Consequences for Residential Green Areas: A Historical Survey of Planning Concepts and Attitudes Towards Green Areas in Housing Estates in Oslo

Jon Guttu

Planning history can be traced to different models of thought, each with significant consequences for residential green areas. Shifts in the thinking about building structure will in its turn set limits for the spatial structure in housing areas and hence for the green structure. Historical examples from the city of Oslo clearly demonstrate this. I am particularly interested in the radical shifts in ideology that characterized the planning profession, and the sociological background of these changes as expressed in the design of social housing, that is to say, in estates planned by housing cooperatives or public authorities for unknown residents of moderate means.

A theoretical entry point for understanding the history of planning and architectural thinking is Françoise Choay's analysis of 19th-century planning regimes (Choay 1970). She identifies what she terms 'critical planning', a response to the 19th-century *laissez-faire* political regime that brought miserable living conditions in the disorderly unplanned towns of that era. She locates three models or schools within the critical movement, each with its own particular approach to solving the problem of poor living conditions. What Choay calls *regularization* sought to improve the existing city by a purely schematic regulation and a new infrastructure. Its main application was as a strategy of public administration. The *progressist* model encouraged the exploitation of technological and scientific innovations, and adopted the methods of industry to defining and solving problems. They searched for new solutions to city problems. The *culturalist* model saw the town in terms of a cultural milieu. Culturists, taking their cue from the humanities, art and cultural debates, were interested in how people experienced the town psychologically and socially, primarily from a historical, retrospective vantage point. Their ideal was the tightly knit rural community of the pre-industrial past. Planning of green areas was a central element in all the three schools, but, as we shall illustrate, in quite different ways.

Figure 6.1 The overcrowded 19th-century city: working class neighbourhoods in Oslo from about 1890

Source: Arbeiderbevegelsens arkiv og bibliotek.

The Swedish researcher Björn Linn (1974) traced the development of the three schools into the 20th century in his PhD thesis. The culturalist and progressist models gradually gained the ascendency in urban development, Linn maintains, first in the shape of culturalism, the garden city movement, and emergence of the

THE CRITICAL PLANNING

Divided into three different strategies

REGULARIST SCHOOL	CULTURALIST SCHOOL	PROGRESSIST SCHOOL
HAUSSMANN	**SITTE**	**THE UTOPIANS**
BAUMEISTER	**UNWIN**	**GARNIER**
STÜBBEN	**SCHUMACHER**	**LE CORBUSIER**
	BRINCKMANN	

Figure 6.2 Different strategies for 'critical planning'

storgårdskvartalet[1] in the early 1900s. From 1930, functionalism gave preference to the progressist model. Together with post-war modernism, the model maintained hegemony until 1970. If we pick up where Linn's history broke off, we can observe 20 or 30 years of a culturalist *mentalité* before modernist ideas gradually took the lead in the 1990s.

Choay's and Linn's theories provide the setting for my choice of historical periodization. The different transitions occur with great accuracy in the Western world, not least in Norway's capital.[2] The various paradigms and transitions also need to be seen as examples of integrated design in which architecture, layout and landscape are brought together. Of course, there city building has been influenced by local tradition, politics and economy in each period. Nevertheless, the common planning ideology is easy to identify through the prism of layout and design.

I shall concentrate on spatial structure and outdoor areas in residential estates. There are two factors in particular that define the attributes of the outdoor area: building structure and density. Layout (including facilities) and design of the outdoor area is an additional factor. All three factors have varied significantly over time.

Apartment Buildings, 1850–1900

It is no accident that we can date the very beginning of social housing in Norway to 1851. It was the year the Arbeiderboligen 1851 (The Working Man's Home

---

1 Swedish: s*torgårdskvartalet*, that is, a quarter with one or more apartment buildings arranged around a central court or atrium and flanked on all sides by streets. It is a basic urban grid element in other words.

2 There are exceptions. Functionalism did not achieve a comparable breakthrough in 1930s Great Britain; the Soviet Union, after being in the vanguard of the progressist movement in the 1920s, met with the politically determined culturalism (social realism) of the 1930s and 1940s.

Association), a bourgeois philanthropic entity, was founded with a mission to improve the housing situation of the working class in Oslo. The run-up to the initiative was dramatic enough. The city had just emerged from a cholera epidemic and a popular uprising led by Marcus Thrane.[3] Both the uprising and epidemic was blamed on the miserable living conditions. A discussion was soon launched which developed into a debate over whether family houses or tenement blocks would be the better response to the situation. The Working Man's Home Association of 1851 supported the latter option. In the years that followed, the association built several estates, of somewhat better design and with lower rents than the private market, and far better than workers could obtain in the suburbs. The number of homes however was too small to exert significant influence on the general housing conditions.

What was the thinking among planners and architects of the time towards the *outdoor area* adjoining these apartment buildings in the Oslo of the 1800s? It is difficult to find any such ideas in the Norwegian debate. We must therefore study the plans and developments themselves. That nothing should be built in the courtyard (or atrium) was one of the principles of the philanthropic projects. As a result, the land was less than optimally exploited. An interesting project was launched by Christiania Selvbyggerforening (Christiania Self-Build Association)[4] in 1878. It displays a 19th-century grid structure, with a relatively large, open space. This courtyard is divided by an access road and the area between the road and building is designated 'garden' (*have*) and 'courtyard' (*gaardsrum*). True, the latter is further divided into smaller sections by outdoor toilets and small storage buildings. Although only half of the apartment complex was built, similar block structures were built a generation later at Lindern, for example (Jensen 1981).

In practice, children played and people mixed in public spaces. There was little traffic in the streets, and as the city was built bit by bit; greens and unbuilt plots were left as open neighbourhood green areas. Parks were laid out, and squares and school playing grounds became features of regularist urban planning. The most successful complex in Oslo's working class districts is probably Birkelunden, today a national heritage site.

Garden City, 1910–1930

Around 1900, the garden city rose to prominence as a solution to the housing question. England was in the forefront, and Ebenezer Howard was its great exponent and organizer. However, with regard to the physical design of buildings and urban spaces, Raymond Unwin was the key figure. He explored principles and practical execution in detail in *City Planning in Practice* (1907). The medieval city, with its irregular forms and 'picturesque beauty' was one of the principal sources of inspiration. In this sense, he took up the philosophy of Camillo Sitte and other

3 Thrane was a pioneer within the workers' union movement at that time.

4 Christiana was the earlier name for Oslo.

Figure 6.3 'Regularization': The reform block in Oslo 1878 (just partly realized)

Source: Drawing by permission of Oslo City Archives.

'culturalists'. But Unwin invested the medieval city with the rationality of the new age; the garden city has to work for modern citizens. An analysis of the garden city's outdoor areas shows clearly marked out spaces, from the private to the semi-private, from the semi-public to the fully public. The continuous rows of houses cut the estates into segments, with long back gardens acting as the private sphere. The buildings are kept low intentionally, consisting generally of only two floors.

Privately organized housing construction stopped abruptly with the dramatic economic collapse of 1899. It would take 30 years before private capital was again willing to invest in residential construction. 1911 is regarded as the starting point of social housing in Norway. At the culmination of a period characterized by acute housing shortages and with a rapidly expanding labour movement, the city council in Oslo agreed to pledge guarantees for loans. At the same time, a housing inspectorate, or housing office as it was renamed, was set up in the capital, in response to calls by architect Harald Hals, who was also the inspectorate's first secretary. Hals had worked in London and the US, and was an avid supporter of the garden city (Hals 2011). True, no garden cities as Howard defined them were actually built in Oslo, only what one could call garden suburbs, but the planning principles were consistent with the garden city idea. First, density was radically lower than anyone had dreamt of in the 19th-century working-class areas. The political objective was to offer members of the working class a home of their own in healthy, rural surroundings, made practicable by new means of public transport like the commuter train and tram-car (Hals 1920). That the Norwegian garden city movement was baptized 'Your own home' (*Egne hjem*), indicated the importance of the ownership structure in addition to the physical concept. An idea popular among the upper classes was that home ownership and land cultivation would prevent further radicalizing of the working class. And the agriculture products, 'the fruits of the earth', would be an extra ration in difficult times (Kjeldstadli 1990).

If we take a closer look at building patterns and spatial structure, we will discover an interesting distinction. While Unwin consistently sought ground contact for every single home in the shape of vertically divided homes, the Norwegian version is far more ambivalent. Ullevål Hageby (*hageby* means garden city) is the best known garden suburb in the country, with a preponderance of horizontally divided homes and free-standing small blocks of flats. In consequence, the private outdoor space and home are separated, and a core element of the English version, the private garden, is lost. The concept of an outdoor private life was not well developed in Norway in the early 1900s; the reason for outdoor lots at the time was to grow crops, and that could be done collectively. In addition, by opting for blocks of flats, it was possible to reduce the size of the living unit. Horizontal division was also less costly. Another point is that detached housing is strongly embedded in the Norwegian ethos. The free-standing small blocks in Ullevål Hageby may be seen as a reflection of this tradition. A consequence of horizontal division is however the loss of the 'long' private garden, a core feature of Unwin's design.

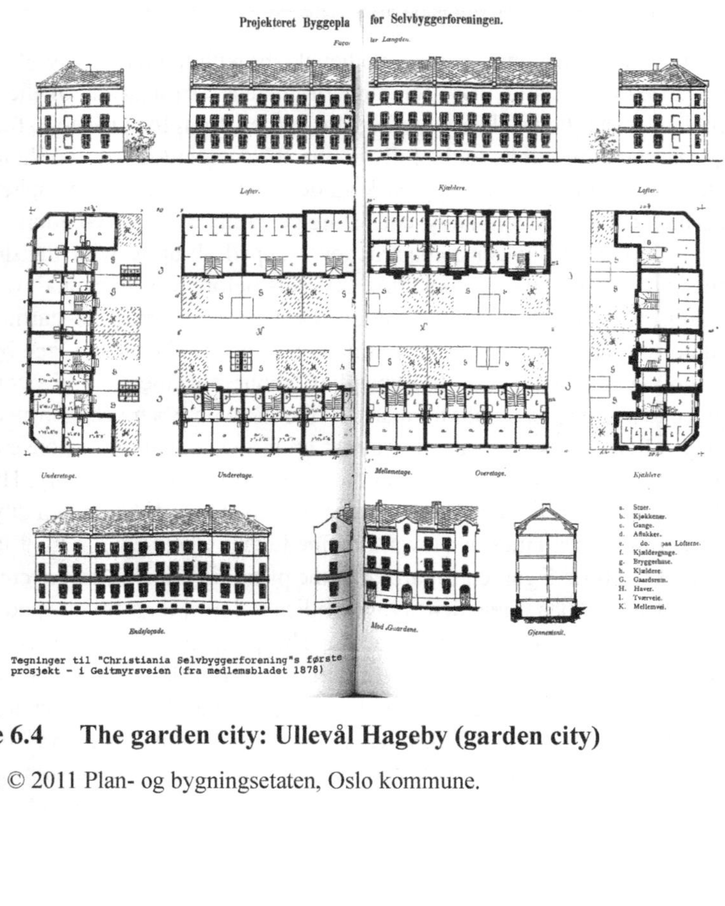

Figure 6.4 The garden city: Ullevål Hageby (garden city)

Source: © 2011 Plan- og bygningsetaten, Oslo kommune.

The garden cities that were built in the Oslo area during the 1920s were otherwise true to the *Art Nouveau* or *Jugend* style regulatory principles, with gently curving roads and buildings that traced the road's curvature. A modest front garden provides a space between the front door and street. We also find in Ullevål Hageby the magnificent Damplassen (Dam Square), a central feature of the area whose buildings are neo-baroque in style and arranged symmetrically around a water feature (a large square pond). According to Hals himself, the inspiration came from his own childhood in the fortress town of Fredrikstad, an example of culturalist-minded housing planners finding inspiration in the past.

Grid Structure, 1915–1930

Financially, Ullevål Hageby was a disaster. The city council had to bear more than two-thirds of the costs. Nor were its residents from the working class, which the original had intended. Rents and deposits were simply too high for ordinary workers. Harald Hals continued his work however from 1918 as the city council's director of housing, and from 1926 as the planning director of Oslo. In the aftermath of the economic calamity of the Ullevål project, the garden city was shelved as a housing option, and the three-story apartment block took over. Nonetheless, Hals continued

Figure 6.5 The courtyard block 1920: Torshov

Source: © 2011 Plan- og bygningsetaten, Oslo kommune.

also to promote culturalist principles in urban planning, primarily through the planning and building of housing estates at Lindern, Torshov and Ila in Oslo.

In this sense, Hals followed the prevailing planning philosophies. Around the turn of the century, several German theoreticians were openly critical of the unplanned housing estate, and launched a new set of principles. They proposed a building structure imposed on the grid structure of the 1800s; the courtyards would be devoid of other built structures, and planted instead with greenery. Rather than rows of buildings comprising small units on their own piece of land, typical of rental housing, the new ownership structure required thinking on a grander scale. Large-scale complexes would be planned and managed by a single property developer, a large company or the town council (Kjeldstadli 1990).

Torshov, planned by the city's planning office, is the best-known example of this approach in Oslo. The architectural ideas as we can see them in the facades still stem from the past, following in an essentially baroque style. Outside spaces, on the other hand, are planned according to new principles. They form a clear hierarchy, from the quadrangular space or courtyards at the lowest level to the city parks at the top. And to amplify the significance of the courtyard, entrances to the complexes faced the courtyard, via an archway. The once miserable backyard was thus transformed into a pleasant park for the enjoyment of the residents.

The complex could be virtually shut off from the surroundings by the archway. The streets were widened, and inside streets, green areas were included in the front of the buildings, in all likelihood to allow sunlight into the windows and embellish the streets. Torshov Park was situated in the centre of the development, on an elevation, and equipped with a music pavilion as if to highlight the value of culture. As the ultimate level of the outdoor hierarchy, expansive green areas were made in the Torshov valley, with connections to other urban areas and the forests surrounding the capital.

Parallel Low-rise Blocks, 1930–1940

Towards the end of the 1920s, functionalism began to make headway in Norway. The Stockholm Exhibition in 1930 marked the definitive Nordic breakthrough. Behind this success were changes in mentality, working methods and architectural style. Perhaps the most ground-breaking and clearest example, of relevance to housing and building patterns, is Tony Garnier's 1904 idea for *une cité industrielle*. Impressively prescient and creative, his vision embraced most of the urban planning principles of functionalism, such as zoning or spatial differentiation by function, continuous ground surfaces and free-standing buildings which, with their polished, unornamented surfaces, distanced themselves from all contemporary styles (Linn 1974).

According to Choay's and Linn's scheme, the progressist model offered a practical approach during the 1920s. This was based on function-analysis, adopted from German and American industry. In France, Le Corbusier gave the new philosophy an architectural design. The city, he maintained, needed root and branch renewal. The 'corridor street', and with it the grid structure, would have to go, to make way for high-rise, free-standing tower blocks raised on columns in an unbroken green landscape. The greens should mainly serve as areas for recreation and pedestrian traffic (Rådberg 1997).

Functionalism in Norway flagged the same principles as elsewhere in Europe. The neoclassical estates of planning director Hals attracted criticism. They did not let in enough sunlight, they were impractical, expensive to build and sagged beneath the weight of anachronistic symbolism (Plan 3 1935). The municipality, which had supplied the organic housing estates of the 1920s, reduced its entrepreneurial activities around 1930, leaving the stage to the housing cooperatives and private sector developers. The Johnsen brothers' firm was the most well known of these

Figure 6.6 Free-standing, low-rise blocks 1930: Sinsenbyen
Source: © 2011 Plan- og bygningsetaten, Oslo kommune.

construction businesses. They were belt-tightening years, and projects had to be economically sound and without unnecessary frills (Ibid.).

The Sinsen estate was the largest of the three brothers' achievements. Some of functionalism's principles are evident in this project. The free-standing buildings are four and five storeys high and with continuous outside areas. The areas are without any clear hierarchy, and with rather a diffuse invitation to use. For economic reasons, the land was exploited to the last square foot, and no larger areas were set aside for outdoor recreation or play within the estate. In that sense, the Sinsen development satisfied only some of functionalism's criteria.

Low-rise Blocks in an 'Open, Park-friendly Urban Plan' 1945–1960

The early post-war years were a time of enormous optimism, but little ready money was available. The hard times of the 1930s were over, and with the social democrats in power, a new society based on equality was to be built. At the same time, after five years of war there was a craving for the safe and sure, the homespun and the national. International trends were less important than national interests, focusing on the need to roll up your sleeves, get your hands dirty and build the country.

These circumstances saw a growing divergence develop between what one might call the 'social' and 'international' functionalists in Norway (Guttu 2003). The social functionalists wanted to reconsider the doctrine of the 1930s. Now the social sciences rather than technology were the place to turn for advice and new knowledge. A question put forward in the Western world was how the organization of the city could strengthen democratic attitudes among people? An answer was found in the concept of the neighbourhood unit. The city should be organized as a hierarchy in three levels, with residential units, neighbourhoods and metropolitan districts. The background for this idea came from garden-city planning together with the works of the influential American historian and writer Lewis Mumford. In his book *The Culture of Cities*, Mumford maintained that cites had grown too large and complex, with isolated and lonely people as a result. Not only was modern city life negative for individuals, it even threatened a democratic evolvement. The neighbourhood should be an arena for communication; the small scale would guarantee a prosperity and human development (Mumford 1970). In Norway, this was combined with establishing housing cooperatives with the residents as self-organized owners. The period was also attentive to the needs of children. The children were the future, and child psychologists recommended play areas in green and safe surroundings. In the atmosphere just after the war, such ideas enjoyed academic hegemony. Functionalism's early orthodoxy had been put to one side.

Small-scale, low density developments were popular; in Oslo this meant low-rise blocks and four-unit apartment houses. One thing the city had enough of was land. An enlargement of the city in 1948, when the municipality of Aker was incorporated within the city limits, made it possible to build low-density housing on farming land. There was little motorized traffic at this point, mainly because car purchase was rationed, and the road system was kept cheap and simple. To reduce costs, the green areas were finished by the residents. Groups of trees were planted and grass seed sown over large areas. But some of the defining characteristics of the 1930s were retained: land-use zoning, free-standing blocks and continuous outdoor areas. As a result, these complexes may be characterized as a mix of culturalist and progressist characteristics. Small scale, low density and vernacular

Figure 6.7 Low-rise blocks in landscaped plan 1950: Keyserløkka
Source: © 2011 Plan- og bygningsetaten, Oslo kommune.

architecture were features linked to the cultural tradition, while land-use zoning, free-standing blocks and continuous outdoor areas were characteristics of functionalism ('the progressist model').

One of the most remarkable participants of the post-war housing debates was architect Frode Rinnan. The greatest innovation since the last war, he said in 1954, is the three- and four-room standard in an *open, park-friendly urban plan* (OBOS 1954). He was thinking of areas like Keyserløkka, Valle-Hovin, Teisen and somewhat later Lambertseter. And it may be here with these developments that one comes closest to functionalism's ideal of buildings situated in a continuous green landscape.

Large-scale Blocks and Cars for the Masses 1960–1970

The late 1950s and early 1960s witnessed another transition. Housing shortages could now be solved in partnership with industry. This may explain the resurgence of the international modernists. Their aesthetic was based on large-scale serial production, modules and installation engineering. The industrial-scale manufacturing of homes meant identical components, apartment types and buildings, built on a grand scale. After a long period with discussions and refusals during the 1950s,[5] the tower block gained acceptance as family homes, and in the 1960s became part of the landscape in many new estates.

At the same time, import restrictions on cars were repealed. Between 1960 and 1965, the number of cars on Norwegian roads grew by 84 per cent. Motoring and the consequent need for parking spaces made land effectively scarcer, and increased competition for land use. Traffic accidents grew exponentially, resulting in debates over planning concepts for outdoor areas. This, in turn, spotlighted the child. Until now, outdoor areas had been accepted more or less as a result of the building lay-out. Children could play anywhere without putting themselves at risk. From now on, play areas should be considered as an integral part of the plan. The most important thing would be to let one side of the block satisfy access requirements, and the other side be set aside for play areas and gather points (Gabrielsen 1965). Traffic would have to be differentiated, with a more precise categorization of types of traffic and speeds. These factors contributed to the emergence of road and outdoor area standards and norms. The standard on outdoor space for children was defined as net 15 m^2 per 100 m^2 floor space. Play areas were divided into sand-pits for the youngest, play areas per estate unit and areas for ball games. Swedish studies of children's behaviour suggested that the children's zone should be located in the continuous non-traffic zone of the estate with footpaths, entrances and designated play areas (Wohlin 1961). It would clearly be unreasonable to expect children to remain inside the traditionally enclosed play area. By implication, the realm of the car would have to be restricted. The amount of attention paid to children's needs stood in strange contrast to what was actually built. The image of the little children on the ground, with their mother on the 13th floor, was the visual expression of this paradox.

It is not easy to describe the spatial structure of the estates of the 1960s. Estates were composed more than before of different block styles and the planners were more concerned with volumes and sculptural construction *en masse* than with shaping urban green spaces for defined uses. Buildings could be long, curving low-rise blocks, to offset the tower blocks, often built on an elevation. And the buildings could also stand in contrast to the wider landscape and green structure.

Ammerud, an estate in the Groruddalen area of Oslo, displays these characteristics in full. A long, curving 'banana block' consisting of four 13-storey tower blocks, each with 250 apartments, produced an image of a carpet of atrium-type homes in great swirls. The wider landscape delineates the estate on three sides. Here, too, the outdoor area makes up a continuous surface, and green areas dominate across the estate. From a distance, attention is drawn to the tower blocks. We are as close to Le Corbusier's vision as we ever have been in Norway.

5 Characterized by questions about fire restrictions, costs and fitness for families with children.

Figure 6.8 Large scale blocks 1967: Ammerud

Source: © 2011 Plan- og bygningsetaten, Oslo kommune.

Small, Dense and Low, 1970–1980

In the latter half of the 1960s, the modernist housing concept came under heavy fire. Inspired by the 1968 revolt, a new generation of planners led the campaign, allied with forces characterized by cultural conservatism (Guttu 2003). In the course of a few years, professional attitudes towards housing quality were totally upended. The large, sculpturally organized building plans were put to one side, the tower block excommunicated. Prefabrication and standardization stopped. The planners looked back to older wooden architecture, national traditions, small family homes and environmentally sound building concepts. Attention was once more focused on social life in village communities with focus on green spaces. The winds of change blew in the direction of culturalism.

Criticism of the suburban large-scale satellite towns came from both the political left and right. But in Oslo, the 1975 local elections handed the conservatives a majority, and they adopted a new housing policy. In future, at least half of the homes built were to be family-size detached or semi-detached homes.

To explain the shift in policy, one needs to look at the demographics. Like in other Western cities with large-scale post-war suburbanization, Oslo had been haemorrhaging young, upwardly mobile families to neighbouring municipalities with family-friendly housing. By giving precedence to the construction of the family home, the city council was offering what this group wanted, seeking to attract valuable taxpayers back to a city with large infrastructure expenditures (Hansen and Guttu 2000).

The idea behind the large-scale family home estate came from Denmark, where it was called 'dense and low' (*tæt lav*). In the 1960s, the talk had been of large tapestries of family-sized homes in rigid patterns (Statens byggeforskningsinstitut 1971). But the upheaval of the late 1960s encouraged or paralleled interest in estate plans allowed for more variation, with curving streets, squares and parks redolent of the Danish village. Sweden called the phenomenon the 'small house surge' (*småhusvågen*). And in Norway, the 1970s became the decade of the densely built family home estate. In many ways, the concept as well as the final outcome can be seen as a revival of the garden city.

So how did planners now conceptualize the good residential area and their green spaces? At the time, *environment* was the big buzzword. As academics and professionals sought to make the most of the concept, one found it necessary to differentiate between 'natural environment', 'built environment', 'social environment' and so on. The planning theoretician and architect Christopher Alexander had formulated the slogan *community and privacy* to underline the balance between two different aspects (Alexander 1963). While the planners of the 1960s concentrated on the private outdoor space, creating the atrium house and stepped house as an answer to the need for privacy, attention was now turned towards the communal features of the estates. Designers wanted to facilitate and encourage neighbour and community contact and a certain level of social control. Common courtyards became the design answer to this, and the prevalent unit of planning. Even the tower blocks at Romsås were built in car-free courts to encourage social intercourse and allow children at play to be supervised. Indeed, communal green areas were given considerable attention because of the children. The play area hierarchy developed in the 1960s was refined further. Motorized traffic was minutely segregated. Two consequences often overlooked were the difficulty of gaining one's bearings within the estate, and optimistically long distances between parking facilities and home.

The family home estate at Holmlia, south of Oslo, is an example of this concept at work. Aesthetically, the planners wanted to hide the buildings in the landscape as far as possible. Buildings were erected on natural land, frequently craggy, environmentally sensitive areas in pine or mixed forests. The ideal was a gabled roof that could be glimpsed through the pine trees. This in contrast to estates of the 1960s, whose structures were often cubic and the colours light in order to differentiate housing from the landscape. To create as strong an impression of the natural environment as possible and downplay the man-made estate, green buffer zones were designed between the pathways and buildings. It was a land consuming layout and required very low-density levels.

The project Hallagerbakken is situated in a steep hill-side. The planners emphasized the quality of the common outdoor space between the terraced houses with the aim to create flat and useable play grounds for children and small

Figure 6.9 Small, dense and low 1980: Hallagerbakken

Source: © 2011 Plan- og bygningsetaten, Oslo kommune.

meeting places for adults. The houses themselves were used to modify the terrain. Encroachment in the existing terrain and vegetation was reduced to a minimum.

Back to the Town and Back to the Grid 1980–1990

Planners not only looked back to the tradition of small family homes, they also looked at the traditional, dense and varied cityscapes. Here, organized city renewal was the door opener. The reasoning was the same as for the family home estate. The central city areas were considered unfit for children, and families joined the exodus to the suburbs as soon as they could. The city centre, its housing stock and outdoor spaces was in urgent need of renewal. Planners looked to Germany, especially Berlin, where the public authorities had long been engaged in renewal and rebuilding in the old tenement districts. The Second World War had fragmented and devastated the urban structure, and made it possible to think in terms of a grand project (Krier 1975).

A strong culturalist ethos was evident in planning circles. Björn Linn's book on '*storgårdskvarteret*'[6] praised the pre-functionalist town and the grid unit as

6 See footnote 1 above.

the building block of urban planning (Linn 1974). There was significant support for conservation. 'Back to the town' proved a winning slogan in political and planning circles and with the general public, and also laid the ground for extensive gentrification (Tone Huse 2010).

What was it the planners and other professionals wished to go back to? First, the enclosed urban spaces of the 19th-century town found support among planners whatever their philosophical viewpoint. While Linn's argument was based on saving land, Christian Norberg Schulz saw the site-specific properties of urban space as important to mankind's existential meaning (Norberg Schulz 1975). The Danish architect Jan Gehl published studies of life *between* the buildings; just looking at other people was the main activity in the public domain, he claimed. Well-defined urban spaces would attract more people, rather than spreading thinly across an open town (Gehl 1971). This then provided the rationale for letting the existing urban structure with streets defining the grid units form the basis of urban renewal.

Second, planners wanted to reinvigorate differentiated land use. Criticism of the zoning, formulated by Jane Jacobs in the 1960s, was now supported by experts as well as politicians (1961). The mandate became: away with functionalism's separation of commerce, manufacturing, homes, institutions and services, and in with multi-purpose and intensive land use. Multi-purpose usage in high-density areas required more careful planning of outdoor areas, in order for example to avoid traffic accidents or to ensure quiet outdoor meeting places.

Third, there was the revival of interest in the grid structure. It became literally the building block of urban development, whether brown field or green field. One of grid structure's advantages was the 'automatic' creation of a hierarchy of green spaces, a distinction between the park and the public street on the one hand and a courtyard, under semi-public and private control, on the other. The grid unit could therefore provide for a quiet, secluded and safe outdoor area, but in the midst of a lively city. The grid unit could be further broken down into smaller zones for different purposes (Linn 1974).

Grønlands Torg (*torg* means market square) was rebuilt in the latter stages of the urban renewal programme, in the years before 1990. The area had previously housed butchers and grocers. It tells us how planners went about putting the intentions described above into practice in the middle of the city. A narrow, diagonal street separates two housing complexes and opens onto a small square. The square itself is divided into different zones. The layout of Grønlands Torg is rather complex. Some activities are at street level, others below ground on either side of an underground passageway leading to the tube. Parking space is located beneath the housing complexes. The two courtyards are completely insulated from the hectic bustle of the surrounding streets. Lots of greenery in the atriums offers residents a peaceful retreat. The consequence for the existing adjacent street however, namely to be shaded and overseen by buildings up to ten storeys tall, is less pleasant.

Figure 6.10 Back to the courtyard block 1990: Grønlands torg

Source: © 2011 Plan- og bygningsetaten, Oslo kommune.

Densification and Back to Regularization? 1990–present

Around 1990, the deepest crisis since 1899 hit the housing industry in Norway. Not until about ten years later would the rate of housing production reach levels of the 1980s. One of the outcomes of the crisis was that Oslo's urban renewal programme went bankrupt. Another was the hegemony of private sector property developers and contractors over public planning in the capital, with the city council's planning agency assuming a back-seat, control function. Oslo had followed the same track as other European cities and British cities in particular, in the wake of Thatcherism (Hall 2002).

In 1987, the Brundtland Report, 'Our Common Future', put renewed focus on the environmental status of the world (World Commission 1987). The report represented a breakthrough for the notion of sustainable development, and this was adopted for cities under the term 'sustainable cities', or 'sustainable urban development'. Coinciding with this, it was increasingly argued that urban densification was the most environmentally friendly urban development option (Naess 1992). Energy consumption for transport purposes would be lower with densification than by enlarging the city and it would also strengthen the possibilities for public transport.

Densification would make it possible to use existing infrastructure and spare virgin areas around the town from development. From 1989, densification became the basis of Oslo's urban development policy (Hansen and Guttu 2000).

Potential home buyers showed that they had similar preferences. In the midst of the crisis, while areas in the periphery were being shunned by buyers, centrally located city accommodation was going for a high price. Living in town had become a popular option for the nature-loving Norwegian, especially for singles.

Over the last 20 years, the population of Oslo municipality has grown by more than 100,000. The transformation of the city is now the accepted shape of development. Property developers are scouring the city for brown field sites or for 'service industries' poised to relocate. Developments are based entirely on commercial principles, with prices determined by the market. High land costs and the hunt for maximum profit add to the pressure, naturally enough, to increase density and build higher. The consequences for outdoor areas are rather dramatic.

While architecture increasingly reflects a modernist approach, the green areas are far from the models of Le Corbusier. In many ways, the planning process and built environment share some of the traits of those of the 1930s as shown in the Sinsen example, just denser. Under cover of the buzzword 'urbanism', tightly packed eight to ten free-standing storey blocks in a schematic lay-out are now quite typical. Shared outdoor areas are continuous, but narrow and rather dark. The public areas of the city and the private balcony are, in practice, the outdoor choices available to residents. When there are insufficient semi-public outdoor spaces, the victims again are the children and less mobile sections of the population. Of course, this has been reflected in the planning discourse in Oslo during recent years.

The Grønlandskvartalene offers a clear example of this development and can be studied as an interesting counterexample to the neighbour complex Grønlands Torg when it comes to building structure and organization of the green areas. While Grønlands Torg was built in two quarters around inner courtyards, the Grønlandskvartalene is built with free-standing lamella blocks and outdoor spaces around and between the buildings. The density in Grønlandskvartalene is about TU 300 per cent,[7] or two times the density of the apartment buildings from the late 19th century. The consequences can be seen in the form of significant differences in sunlight and ambient lighting within the estate. The square called *Teaterplassen* offers a breathing space, providing functions other than accommodation. But it is hardly suitable as a playground.

Conclusion

One hundred and fifty years of planned housebuilding in the city bears witness to great changes in the way planners think and to widely differing results. Let

7 TU=300% is equivalent to a floor space index about 2.7.

Figure 6.11 Concentrated free-standing blocks 2000: Grønlandskvartalene

Source: © 2011 Plan- og bygningsetaten, Oslo kommune.

us now return to Choay's theory on critical planning and the three models of thought, and discuss whether it is meaningful to apply this theory to the material I have presented. Critical planning arose, as mentioned, as a reaction to the conditions in the industrial cities of the 19th century, not least to the lack of green areas. In spite of significant changes in city structure and city life, it is my hypothesis that the planning models identified by Choay have been influential throughout the 20th century with significant consequences for outdoor areas.

First, we can state that the different examples clearly express shifts in planning ideology, mainly in accordance with the scheme of Choay. One can count four essential changes within the period:

1. From the regularist period in the last 19th century to the culturalist era in the first three decenniums of the 20th.
2. The breakthrough of functionalism (progressivism) about 1930.
3. The shift from progressivism to culturalism round 1970.
4. Back to regularization after the economic crises in 1990.

The first three of these shifts represented radical breaks with the practice of the time: they had an ideological as well as a formal content, and were easily identifiable. If, on the other hand, we move on to the turn of the century, it is more difficult to identify any one example as representative of a particular model or school of thought. It seems as if both progressivism and culturalism are empty of ideological and formal content, although the architecture is characterized as neo-modernism (Grauer 1981/1982). Are we back in a model of regularization, where large building companies transform low exploited areas, while the authorities organize the necessary infrastructure and then use regulatory tools to influence a more or less schematic concept? There are interesting similarities between the present situation and the late 19th century.

If we look at the examples in detail, we will see that several of them contain features which cannot be directly traced back to any one of the three models mentioned. We may say that the projects have been modified in the face of the ideals of the times. There can be several reasons for these types of modifications, the most important of which is undoubtedly where the complex is located in the city. This will affect the density and hence the green structure. High-rise flats with landscaped outdoor areas, like Le Corbusier dreamed of in the 1920s, only became feasible on a mass scale after World War II, owing to the supply of land when the city expanded and built satellite towns. In addition, construction methods, building regulations as well as cost of installing lifts played their part. The examples Sinsen and Ammerud show this. While Sinsen was developed with blocks of flats up to five stories high, the satellite town Ammerud could realize several properties of the progressist model.

Furthermore, the progressist examples show that the residential green areas more or less can be seen as a consequence of the building structure. This is because of the starting point, namely that buildings are put into an untouched landscape. This scheme may function when one builds in the periphery with moderate density. The same concept applied in a dense situation will, like we can see in the Grønlandskvartalene, lead to narrow and dark residential green spaces with insufficient public control.

In the culturalist examples such as Ullevål Hageby, Torshov, Hallagerbakken and Grønlands torg, on the other hand, the green space plays an independent and important role in the planning process. It is my opinion that to formulate the intrinsic values of outdoor spaces can help to protect green values in planning.

One hundred and fifty years of planned housing construction in the city displays significant changes in the mentality and philosophy of the planners and in the practical outcomes. We have noticed huge variations in building structure and density and significant consequences regarding outdoor spaces. These transitions can be understood as an oscillation between backward-looking and modernistic philosophies, both of which can result in well-planned outdoor spaces, even though they operate on different premises.

Bibliography

Chermayeff, S. and Alexander, C. 1965. *Community and Privacy: Toward a New Architecture of Humanism*. New York: Anchor Books.

Choay, F. 1970. *The Modern City: Planning in the 19th Century*. London: Studio Vista.

Gabrielsen, E. 1965. Våre omgivelser utendørs – en vesentlig del av vår bostandard, Bo, 2(1), 12–16.

Gehl, J. 1971. *Livet mellem husene: Udeaktiviteter og udemiljøer*. 1st Edition. København: Arkitektens forlag.

Grauer, V.A. *Downtown Review*, 3 Nos 1 & 2, Fall/Winter/Spring 1981/82.

Guttu, J. 2003. *Den gode boligen: Fagfolks oppfatning av boligkvalitet gjennom 50 år*. Oslo: AHO Avhandling 11.

Hall, P. 2002. *Cities of Tomorrow*. Oxford: Blackwell.

Hals, H. *I Store norske leksikon*. Available at: http://snl.no/.nbl_biografi/Harald_Hals/utdypning_-_1 (accessed 24 October 2011).

Hals, H. 1920. *Hus for smaakaarsfolk: Ti aars boligarbeide i Kristiania.*[8] Christiania: Kristiania Kommune.

Hansen, T. and Guttu, J. 2000. *Oslo kommunes boligpolitikk 1960–1989: Fra storskalabygging til frislepp*. Oslo: NBI and NIBR.

Huse, T. 2010. *Tøyengata – et nyrikt stykke Norge*. Oslo: Flamme forlag.

Jensen, R.H. 1981. *Moderne norsk byplanlegging blir til*. Trondheim: Tapir.

Kjeldstadli, K. 1990. *Oslo bys historie*, 4. Oslo: J.W. Cappelen forlag.

Krier, R. 1975. *Stadtraum: In Theorie und Praxis*. Stuttgart: Karl Krämer Verlag.

Linn, B. 1974. *Storgårdskvarteret: Et bebyggelsemønsters bakgrund och karaktär*. Stockholm: Statens institut för byggnadsforskning.

Mumford, L. 1970. *The Culture of Cities*. San Diego, CA: Harcourt Brace Jovanovich.

Naess, P. 1992. *Natur- og miljøvennlig tettstedsutvikling: Faglig sluttrapport*. Oslo: NIBR-rapport 1992, 2.

Plangruppen. 1935. Byregulering eller konditorkunst. *Plan*, 3(1), 9–14. Oslo: Socialistiske arkitekters forening.

Rådberg, J. 1997. *Drømmen om atlantångaren: Utpier & myter i 1900-talets stadsbyggande.* Stockholm: Atlantis.

Rinnan, F. 1954. Boligstandarden i Oslo 1929–54, in *Oslo Bolig og Sparelag 1929–1954*, edited by Oslo Bolig og Sparelag. Oslo: Oslo Bolig og Sparelag, 43–55.

Unwin, R. 1907. *Town Planning in Practice: An Introduction to the Art of Designing Cities and Suburbs*. London: T. Fisher.

Wohlin, H. 1961. *Barn i Stad: En studie i tre delar av barns utelek och trafiksäkerhet i nya bostadsområden*. Stockholm: KTH.

Verdenskommisjonen for miljø og utvikling. 1987. *Vår felles framtid.* Oslo: Tiden Norsk forlag.

8 Oslo was called Christiania until 1925.

Chapter 7
Green Suburbanisms: Differentiating the Greenness of Suburbs

Per Gunnar Røe

Introduction

Suburbanization and suburbs are generally considered antithetical to an urban sustainability or green urbanism. In fact, insofar as the suburb is linked to urban sprawl and the development of a 'suburban lifestyle' directed toward excessive consumption, the suburb is the antithesis of green urbanism. There is little doubt, indeed, that the building and layering of suburbs in increasing distance from city centres, and the massive building of low-density housing with little or no public transport, particularly in American cities, has contributed to unsustainable development especially with respect to transport emissions and energy use. Furthermore, car dependent lifestyles associated with 'mainstream' (North American) suburbia and suburban living are undoubtedly unsustainable (see Næss 2006, Newman and Kenworthy 1989). Post-war suburban development has, in addition, changed the landscape and led to the destruction of farms, forests, natural habitats and to the denigration of local streams (Hanlon et al. 2010).

Nonetheless, close examination of suburban types and the process of suburbanization reveal a more nuanced picture, spatially as well as socially and culturally. In addition to the privately master-planned middle-class suburb, representing the suburban myth, there are working-class suburbs, immigrant suburbs, poor suburbs, high-rise suburbs and so on. In short, the spatial form, housing types and type of transport infrastructure vary widely. Thus general conclusions about suburbs based on research on urban sustainability may not hold in all cases.

Suburbanization, and suburban history, is thus more differentiated and varied than the dominating representations of 'suburbia' imply. Herbert Gans (1967) was one of the first to deconstruct the myth of suburbia as both homogenous and fundamentally different from urban spaces and urban life. More recently, Harris and Larkham (1999) have pointed out that the form and the social composition of suburbs are highly and increasingly differentiated. Crucially, the discursive simplification of suburbia has led to an inability to 'read' the complexities and problematics of a varied suburban landscape, and a lack of accuracy in how we deal with suburbia within governance and planning. A lack of differentiation has made policies insensitive to the variation in social context and socio-spatial arrangements and practices, including those that counter or support green urbanism. In this chapter I will focus on the green spatiality of suburbs as an aspect of liveability that helps explain why people move to suburbs, but also creates the expectations that have led to urban sprawl.

I argue that it is of importance to look for the green threads in the history of suburbanization and suburbia, and to consider more closely how the variety of suburbanisms have been constructed through discourse and practices within politics, planning and everyday life. Knowledge on how suburbs have been constructed, discursively and materially, may give us valuable insight and a more differentiated and contextually sensitive understanding of the green aspects in an increasingly differentiated suburbia. This insight is of importance for the attempts to retrofit suburbia and in making existing suburbs greener. Part of this policy is building more compact housing and housing areas and densifying built-up areas, which in some cases put pressure on green recreational spaces, reducing the liveability of places.

My intention in this chapter is to discuss how green spatiality, its design and practices, is part of different suburban discourses and socio-spatial relations. I will approach this through representational spaces, to use Lefebvre's (1991) concept, describing professional discourse, knowledge and framing of the suburbs of the modern city. I will consider how the suburban green spaces have been constructed as part of the planning and design of types of modern suburbs, focusing specifically on an outer suburb of Oslo, Skjettenbyen. This suburb may be considered a case of post- or modified modernism in green planning, or a green suburbanism, that serves as a middle ground between the modernist planning discourse, represented by the communitarian utopians and functionalists of the late 19th and early 20th centuries, and a postmodern privatized planning of suburbs represented in its extreme fashion as 'privatopia' by Knox (2008). This case reveals that understanding the tension between density and liveability is crucial for the green transformation of housing and living spaces, that is, to create spaces that are compact and accessible and that permit socio-spatial changes in relation to dwelling, work and recreation.

Picturesque Suburban Green

The massive suburbanization of Western cities was introduced when the upper-middle class, and later the middle classes, fled the environmentally degraded industrial cities of the 19th century; its phases are described by Jackson (1985) and Hayden (2003). An increasing proportion of the wealthy social elite of the time found the smoke and noise in the city intolerable, feared the social unrest of the growing working class and working poor and moved into the green countryside, permanently or during parts of the year (the real upper class maintained city residences and acquired multiple houses in the countryside and resort locations). According to Fishman (1989) this was first evident in London, where the birth of the modern suburb took place in the 18th century. These borderlands and picturesque enclaves (Hayden 2003), often planned and placed in park-like settings, represented an escape into a green haven only accessible for the merchant and professional wealthy elite. At this time, before automobile use became widespread, these developments did not represent an unsustainable transport, but were socially unsustainable.

These 'green field' developments took advantage of the green spaces they were situated in, making large private gardens and landscaping the 'natural' surroundings, and as Knox puts it; '[...] encompassing landscaping that was carefully manipulated to give a natural appearance [...]' (Knox 2008: 16). The suburbs were made for the few and wealthy, and thereby contributed to the social polarization of the time by adding the geographical component of segregation. Many of these suburbs are still wealthy enclaves. One example is Bedford Village, a New York suburb where, according to Duncan and Duncan (2004), the green landscape acts as cultural codes conveying social distinction and hierarchy. The 'natural' and green landscape of these early modern suburbs represented places to which the rich and upper classes could escape. Later the aesthetics of this landscape, with its large lots and low-density housing, has been preserved through zoning and restrictions to prevent the suburb from changing by permitting the entrance of lower social classes. In this case the preservation of green landscape is contributing to social inclusion and exclusion.

Oslo had its share of such developments, along tram (street car) corridors up the hillsides, west of the city. The tram, established in 1905, was to bring the growing number of urbanites into the vast recreational forested areas north of Oslo. However, it also laid the ground for real estate development along the line for fortunate city dwellers who wished to escape the urban noise and squalor. Large lots (of minimum 4,000 m^2) were laid out to attract only wealthy residents, resembling the exclusionary zoning practiced in other upper- to middle-class suburbs. According to Nilsen (1998), the developers in these areas wished to avoid unpleasant establishments such as hospitals, factories and working-class housing, as well as any activities that would cause smoke, noise, smells or other unpleasantries for 'the neighbour' (Nilsen 1998: 52).

This development privileged the recreational preferences and housing culture of the upper classes, providing them with large private lots, direct access to the woods and swift transport to white collar work places in the city, but also produced a suburban green aesthetic. Even if these developments, in Oslo, didn't apply as much cultivation of the landscape as in North American and British cities, the landscape setting, with the woods as a background, provided an aesthetic ideal. The representation of the wealthy suburbanite was set in a spacious green landscape with a mixture of natural and landscaped elements.

Utopian and Modernist Green

The 'rationale' for such social exclusive suburbs is distant from the then contemporary philanthropic and modernist utopianism of the late 1800s and early 1900s which also came to influence the process of suburbanization, the development of the suburban landscape and the social construction of suburbia. While some would escape the industrial city of the late 19th century, others wished to reform it. Factory owners (like Robert Owen), scientists (like Patrick Geddes), writers (like Ebenezer Howard), architects (like Le Corbusier) and others with a more or less philanthropic ambition to change the urban scene developed ideas, projects and plans for new cities and suburbs.

In the context of this chapter it is important to draw a line between the utopians who wished to break loose from existing cities to develop new settlements in the countryside and those who wished to rebuild the city itself. Le Corbusier is the most renowned of the latter, exemplified by his plan (Plan Voisin of 1925) to demolish most of historic Paris north of the river Seine, and replace it with 18 high-rise towers with open spaces in between (Hall 2002). Even if Le Corbusier's vision was not a plan for suburban development, it paradoxically had a great influence on mainly state-led suburbanization, most profoundly the building of new towns and satellite towns in Europe and Scandinavia (including Oslo).

While Le Corbusier's thinking was heavily imbued with the technological developments of the time and his urban visions were based on the leading role of new technologies making cities and the city a centripetal force in society, this was not the case for some of the utopians who wanted to break loose from the city. In the 1920s Frank Lloyd Wright developed the idea of a low-density marriage of suburb and countryside. Calling it 'Broadacre City' (Frampton 1992), Wright's design suggested that everyone could live in green spacious environments, resembling the edges of the sprawling metropolis which developed half a century later. On the one hand, Wright argued that a new system of dispersed anti-urban land settlement should be developed (his book was titled *The Disappearing City*). On the other hand, he stated that there was little need to plan this process since it would happen spontaneously (Frampton 1992: 190).

Wright's ideas may seem to have been realized on the privatized exurban edges of North American cities, not so much because of the numerous suburban houses he designed between 1932 and 1960 (Frampton 1992: 191), but because of the resemblance with 'Broadacre City'. However, they were to a little extent adopted in the professional planning discourse itself. Another utopian, Ebenezer Howard, developed his garden city concept in the early 1900s; he would have a much more profound influence on planned surburbanisms than Wright. Howard's vision – combining the fresh air of the countryside and the job opportunities of the city (Short 1989) with a romantic architectural design inspired by German villages surrounded by a 'green belt' – would inspire future planners. In its implementation it was transformed into garden suburbs, like Hampstead in London, with few urban facilities and work places contributing to urban sprawl, especially in British cities (Whitehand and Carr 1999). Now many of these projects have become inner suburbs in a vast urban region. Where densifications have been restricted, these suburbs appear as green, with relatively large commons, as for example the Hampstead Heath.

Recently Howard's ideas seem to have gained renewed interest, from New Urbanist architects and planners, because of its aesthetics, form and land use principles, and because of the representation of a socially mixed, liveable, self-sufficient and walkable community. However, Howard's communitarian ambition and organizing community life have been left behind by the New Urbanists (if not intentionally), who focus almost exclusively on architecture and layout when shaping out mainly private master-planned communities. The design itself is believed by New Urbanists to create, or at least contribute to the building of, social capital within the new estates, resembling a criticized architectural determinism which cannot be supported by empirical research (see Talen 1999). Even though New Urbanist developments do create communal green spaces and differentiate the public-private divide, they are highly desirable for high-income groups and seldom offer affordable housing (Newman, Beatley and Boyer 2010), and they often fail to contribute to the protection of the natural environment because they are, in many cases, set in green field areas (undeveloped land), and because they are often built in places where residents are reliant on the automobile (Hanlon et al. 2010).

The utopians of the late 19th and early 20th centuries were concerned with the environmental (and social) problems of the industrial city. But unlike the privately planned park-like suburban enclaves of that time, they wanted to set aside green commons and public parks for recreational purposes, as Le Corbusier puts it: 'to promote healthy conditions and create a tranquil atmosphere that will offset the strain produced by the accelerated tempo of modern business' (Guiton 1981: 98). The dense city was to be torn down or kept at a distance. In the garden city concept this included not only picturesque landscaping and beautiful gardens, but also a set of land use and planning principles focused on arrangements to separate dwelling from working, as well as public transport between towns and cities. Combined with the community planning principles and neighbourhood principles of planners like Clarence Stein (Hall 2000), these ideas influenced the modernist planning tradition emerging before and after World War II in Europe. This legacy also influenced the planning of Skjettenbyen, as shall be discussed below.

In Oslo, garden city planning principles became part of a modernist planning and design discourse, with the building of suburban satellite new towns (resembling British new towns, without being self-contained with workplaces and services) in the post-war years as the most profound material manifestations. The Greater Oslo master plan of 1950, which laid out these developments, also included a proposal for a new park system, not only connecting the new suburban towns with the forested and protected areas beyond ('Oslomarka'), but including all types of recreational areas and green paths for walking and skiing between all urban districts and the natural and protected areas in the city surroundings (Lorange and Myhre 1991: 148). This top-down planning achievement established a welfare rationale for green (sub)urbanism: a modern green infrastructure aiming at supplying all inhabitants with access to small green recreational areas, as well as large grounds and forests, the variety of spaces needed for 'the modern man'. There was a hierarchy of outdoor areas, from the private balcony and common grounds close to the blocks of flats, to natural and recreational areas and areas for sport (Thorén 2002). In some of the older satellite towns, like Lambertseter, the original green landscape features, terrain and vegetation were to some degree kept as well. This softened the impression of architectural brutalism that the large building blocks would create.

The modernist planning discourse that was instrumental in shaping urbanism was, I would argue, even more important in directing suburban development, especially in Europe. The movement was strongly inspired by the thoughts and writings of radical architects and planners in the 1920s and 1930s, with the German school of Bauhaus and Le Corbusier's CIAM (International Congresses of Modern Architecture) being the most important examples. These ideas were spread through magazines and exhibitions, as exemplified in the important Stockholm exhibition of 1930. Through these influences, an internationalized modernist design and planning discourse developed. A key element in modernist and functionalist planning was to bring light, air and green areas into the city. It has been given little attention that in most cases this emphasis led to suburbanization, as it did for the garden city concept. It just was not comprehensible (or wise) to tear down and replace the old city, although several cities after World War II demolished part of their core areas (Stockholm is one example) and several others had significant parts of the urban core demolished by bombing during the war, making space for rebuilding according to the modernist architectural and planning discourse. It is important to note that although American cities also had garden city-like development, most notably Radburn in New Jersey, where a system of pedestrian ways was developed to permit passage '[…]through communal open spaces between the houses […]' (Hall 1982: 61), the functionalistic and modernized form of rationalized and master-planned suburbanization that held sway all over Europe was in part different from much of the suburban landscapes developed outside North American cities, which starting in the 19th century had been more heavily influenced by the private housing market.

Privatized and Selective Green

Alongside the modernist planning discourse and practice, transforming the former agricultural and forested surroundings of cities like Oslo in the post-war years, there has been a bottom-up, self-built or company-built suburbanization going on, more or less unplanned by public authorities (or laid out by private developers). This was already evident in the late 19th century of picturesque and landscaped suburbs of the elites in North American and European cities, for example in the exclusive villa suburb of 'Homansbyen' adjacent to the royal palace in Oslo – the first suburb in Scandinavia where houses were built before they were sold. But it was most profound in the post-war era of the motor car and urban sprawl, pushed forward by state economics, policies and ideology. Knox argues that during the middle decades of the 20th century, 'American metropolises were unbound by the combination of increased automobility and the blossoming of egalitarian liberalism in the form of massive federal outlays on highway construction and mortage insurance' (Knox 2008: 37).

In many of these suburbs, not only in North America but also in Australia, the dominant share of green recreational spaces consisted of private gardens. Private gardens attracted people moving out from the city and private gardening, with all its styles and practices, became a part of the suburban lifestyle. This attraction of private domesticity and the wish retreat from the city (where modern suburbanites mainly worked) and to live closer to nature is emphasized in Anglo-American literature on suburbanization and suburban ideals (Harris and Larkham 1999: 10). Common green areas were in many cases minimal. Regarding the famous and enormous Levittown development (by William Levitt and sons) on Long Island, New York, Hayden (2003: 136) points out that the nine swimming pools and the seven small parks with children's playgrounds and some baseball fields was extremely minimal for a city of 80,000 people. Hayden explains this partly due to the lack of a master plan. It was the private garden that became the green haven, a retreat from city life, for the suburbanites. Askew and Guirk view the suburban garden as a cultural product, as it is created from a space which allows for individual expression in its creation and display. The social and lifestyle aspirations (that is, cultural capital) of households are on display through newly constructed gardens and recently constructed homes (Askew and Guirk 2004: 18), and for instance the lawn may be seen as a reflection of good taste, purity and decency, and a possibility for the individual to constitute onself as a good neighbour, industrious individual, a person with developed aesthetics and/or keeper of order (Feagan and Ripmeester 2001). The garden and gardening is saturated with meaning, also because gardening is meaningful work, and the garden an expression of dreams and aspirations. It also is part of an urban culture of taming nature and making it fit into culture, in the same way as green roofs function on city buildings.

In suburbs with a large share of semi-detached or multiple units housing, common or semi-public recreational grounds, playgrounds and parks are more prevalent. In Oslo and its surrounding municipalities, public authorities laid out plans that set aside such areas or fields, adjacent to where private developers and entrepreneurs built houses. Planning regulations that require the inclusion of playgrounds or recreational grounds within new housing areas were, to some extent, also applied to areas dominated by detached housing. In much of the post-World War II period, local public authorities influenced suburbanization through land use planning, and when the privatization of planning gained momentum in the 1990s, the state authorities introduced regulations and guidelines to secure recreational and green areas, mostly aiming at the perceived needs of children and youngsters.

The Social Green Suburbanism of Skjettenbyen

The late 1960s and early 1970s saw a revolt against modernist rationalist planning, as part of a postmodern awakening as Ley (2003) argues. Civil society, activists, architects and social scientists firstly reacted against the top-down planning practices based on nearly universal design principles, and the organization of house building though mass production, with little room for public involvement and user participation. Secondly, they reacted against the large-scale, high-rise and brutalist design, and the arguable social consequences, especially for children who were seen by a new generation of planners and sociologists as potential victims of the lack of environmental variation and visual stimulation in the so-called 'green deserts' between the blocks of flats. In addition, some of these estates had social problems and experienced crime, vandalism and decay, making them unpleasant places to live. Several reports, internationally and in Norway, made the case that the design principles led to social misery, in a deterministic fashion. The criticism of such high-rise estates halted further construction (Brattbakk and Hansen 2004).

The reaction against this type of design, on the one hand, and the fear of the American type suburbanization leading to urban sprawl, on the other, nurtured the work amongst architects and planners around 1970 to develop a middle ground design concept. Inspired by Danish architects who were the forerunners, it was argued that the solution was low-rise, but dense building structures (the so-called *tæt-lav* formula: dense and low). The housing areas should be relatively dense, with higher densities than British garden cities (Guttu 2003), but at the same time each dwelling should have contact with the ground. Skjettenbyen, in the north-eastern outskirts of Oslo, is probably the most known Norwegian example of such housing projects where new design principles were implemented. It was built between 1969 and 1973 and consists of row houses with 1,050 dwellings and low-rise apartment blocks with 600 dwellings, organized as a cooperative with inhabitants as shareholders. Skjettenbyen today has more than 6,000 inhabitants.

As it represents a middle ground between modernist large-scale planning and the privately individualized suburbanization, and between picturesque garden and functionalist modernist layouts, Skjettebyen may be seen as a case of design dealing with the ambivalence and tensions both between public and private, and density and liveability. It is a discursively saturated and heavily planned housing project; unravelling its story reveals how these tensions and ambivalences are made knowledgeable through a planning process, where discourse meets materialities and practices. This is based on data material collected as part of a study of suburbs and suburban development in the Oslo region (Røe 2009a, 2009b). The material consists of planning documents and detailed design descriptions (including a detailed handbook handed out to all residents) and articles from the onset of the Skjettenbyen project, as well as reports and guidelines by Danish researchers developing the 'tæt-lav' discourse. In addition, interviews with inhabitants, planners/architects and real estate agents were carried out. I will first dig into the planning discourse framing this project, the design itself and then stories of how it was received and experienced.

The design of this large-scale housing project was a result of an architectural competition won by a team headed by a Danish architect, Nils-Ole Lund. The buildings, designed as modernist styled cubes with flat roofs, were arranged in groups of rows, with a differentiated system of green areas in between (see Figure 7.1). For every dwelling there should be a playground for small children, of minimum size 50m^2. For every 200th dwelling, and within a distance of 150 m from each dwelling, there was to be a large playground, of minimum size 1,500 m^2. And for every 600th dwelling, there should be a ball field, of minimum size 6,000m^2.

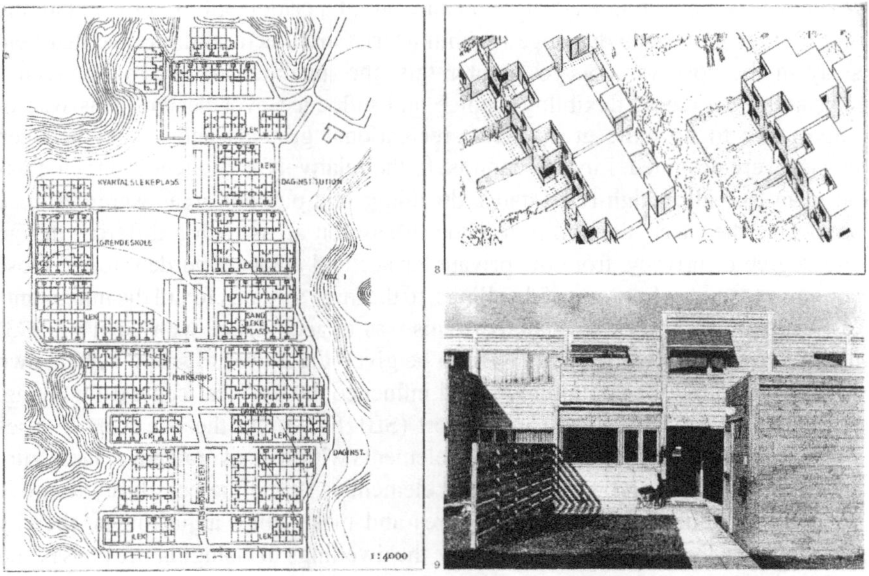

Figure 7.1 Layout and design of Skjettenbyen

Source: Lund, N. 1973 Skjettenbyen, Norge. Arkitekten. No. 10–11.

The project included a plan for a park with pedestrian thoroughfares, and a central playground near the school. One childcare institution, with the capacity for 40 children, was to be established for every 250th dwelling (Dobloug et al. 1972). Other facilities and services were also established: a shopping centre and basic social and health services. There was a hierarchical road system, inspired by the famous Radburn plan, with main roads/thoroughfares passing outside the housing area, and with a distributor road and cul-de-sacs within the area. Cars were to be kept apart from the dwellings, which were to be accessed by foot. Furthermore, the houses were organized so that people could establish contacts across pedestrian tracts (Lund 1973: 187).

The dense and low ideas and principles, which established as a strong discourse especially in Denmark, were partly inspired by an environmental psychology developed by Ingrid Gehl. The living environments of people should satisfy several psychological needs: contact with other people, the possibility for isolation and withdrawal, visual and auditory experiences, activity, play, orientation and structuring, identification and aesthetic stimulation. These needs were to be considered in planning and design, with respect to the following factors: dimensions (height, length and width), arrangements (of benches, trees and play facilities), location (of dwellings in relation to play areas) and sensory stimulus from the environment (Gehl 1971: 167–8). This highly scientific way of designing heavily influenced the Skjettenbyen project.

On the other hand, the Skjettenbyen project incorporated a democratic or participatory element in its design. It came out of a critique of top-down planning and brutalistic designs developed within a strict architectural discourse that held sway in the post-war era. To counter this, the inhabitants were to be given a substantial degree of flexibility, choice and influence over design. This was of importance to how the green areas, recreational grounds and spaces between houses were designed. First, in contrast to the relatively homogenous green areas on many so-called high-rise estates, dwellings and outdoor areas were designed as to give the 'user' the ability to choose between areas with a different design and degree of privacy, from the private terrace and adjacent garden, to the close common grounds of a group of dwellings, to the large grounds for all the inhabitants of Skjettenbyen, as described above. This was in accordance with Gehl's (1973) ideas. Secondly, the inhabitants were to be given the power to shape her/his own environment, and the community could influence the physical context for living, as explained in a Danish research report (SBI 1971). On the one hand, houses were built industrially, of prefabricated elements within an overall plan (see Figure 7.2). On the other hand, the design and elements ensured the possibility to move walls, add building cubes of 3x3 metres and make other adjustments, thereby personalizing the living spaces and make the dwellings fit changing life phases and styles, despite and within a modernist aesthetic and functionalistic design frame.

This structural flexibility also laid the ground, literally, for the green areas. The design principles, manifested in a handbook written by the architects (Dobloug et al. 1972), and handed out to its residents, did not only prescribe how to modify the houses, but also the surroundings and in-between spaces. Lund (1973), the chief architect, explains that the housing system is designed in order to inspire changes, and that the handbook is to show the inhabitants the possibilities in the dwelling system and structure, as well as the technical and economic limitations. The handbook described how cubes could be added or, giving a number of combinations and house designs, each given names to identify the layouts. These designs also influenced the green spaces and microclimate, sun and so on between the buildings.

Not only were the design and maintenance of common grounds and semi-private spaces described in the handbook, but also how private gardens should

Figure 7.2 Views of the newly built project

Source: Lund, N. 1973 Skjettenbyen, Norge. Arkitekten. No. 10–11.

be developed, with plant types, and so on as part of the description (almost half the handbook of over 100 pages deals with the outdoor areas). For the common grounds, the handbook designates the development and maintenance of the three mentioned types of playgrounds and recreational/sport areas. The landscape design of walking paths, and recommended types of plants and trees, were also part of the book handed out to the people moving into Skjettenbyen. In addition, the handbook has prescriptions and recommendations for the small private gardens of the project, how the grounds should be prepared, and what kind of plants would be suitable considering the regional and local climate. There is even a description of how agricultural products and vegetables could be grown on small plots, and under glass cover.

The overall plan, architectural design and the layout for the green spaces were to give the inhabitants contact with 'the ground'. According to Lund (1973), the project aimed at creating a 'socially good' housing area by giving a variety of people, also those with relatively low socio-economic status, good dwellings in contact with the ground, and served with schools, kindergartens, shops, parking places and green areas. Secondly, the future inhabitants (who couldn't participate in the ongoing planning process) were to be given the opportunity to improve their homes, including their outdoor areas. The Skjettenby project was thus created through a planning discourse providing a strict overall plan, structure and programming of the environment, and at the same time organizing a participatory and flexible process of filling in the details. But how is this structural flexibility received amongst people living in Skjettenbyen?

The differentiated and hierarchical structure of green spaces may fit into at least the functionalism of practices that people in certain life phases seem to adhere to, quite in line with how the planners and architects (like Lund 1973) envisaged the housing area. For a woman in her mid-50s, who had raised children there, the green functionalism of Skjettenbyen resonated with an earlier life phase. She emphasized that the playground in front of the houses, and the extensive green space, also suited the children's need to explore '[…] the small woods where they could build a cabin'. Reflecting the general functionalism of the place, another woman said that there seems to be everything you need there: a full range of medical professionals, grocery stores, post offices, schools and several playgrounds in an area with restricted car use, making it very safe for children.

Also real estate agents, who would assess the area qualities and communicate both the symbolic and functional aspects of a housing area (and therefore are of interest when studying the discursive construction of housing qualities), appreciate the way of planning and designing the area. One local real estate agent remarked that these dwellings sold easily, while another stated that the functionality of these houses is appreciated, making them highly desirable. Too few row houses are built, he said, and the ones that are built do not have the same qualities as Skjettenbyen

> […] because they are squeezing it all in, and make it as cheap as possible, and get as many units as possible into the plot they have bought, while things were different in the 1970s and 1980s.

The green commons, and their accessibility, was also appreciated for outdoor recreation. 'We have enough outdoor recreational areas', one interviewed man in his late 60s stated, 'and he added, […] there are pedestrian roads all over, so you can walk several kilometres without crossing a road with cars'. He added that they also had large hiking areas nearby, with tracks for cross-country skiing. '[…] we have everything we need, with regard to this', he said. A mother in her late 30s also emphasized this: 'If you're into hiking, there is nothing to stop you here'.

A planner in his 60s who lived in Skjettenbyen for some years after the opening in 1973 argued that it now would be difficult to plan and set aside such large common recreational grounds, specifically large areas for play, ball games and so on as was done in Skjettenbyen:

> That's the advantage of planning it all from the start. There are no questions. If we now try to set aside space for a ball ground on the plots of private developers, he will cry […] because he then will lose a couple apartments.

The impression that Skjettenbyen was produced in a 'prime time' for planning is supported by a retired architect who moved there at its opening. He testifies to its currency in the new emerging housing and planning discourse among newly educated and radical planners and architects in the 1970s:

> After we had lived in an apartment block in Oslo, we had bought a row house and were to move into Stovner [a large and partly high rise satellite town closer to Oslo]. But then this emerged. So we sold before we moved on, and came here. Another architect I know also moved here [… as well]. This was an exciting and radical project, and young architects liked it.

This architect had been designing social housing, and for him architecture is closely related to political issues. The row houses in Skjettenbyen would fit in, and could serve as symbolic capital, or an expression of his social, as well as his professional identities. In line with the environmental psychology and planning discourse of the 'dense and low' movement, amongst young and progressive architects, large-scale projects were divided into smaller community units to promote an overview and a sense of belonging lacking in large-scale and high-rise projects. In the Skjettenby project the remnants of the natural landscape were meant to contribute to this. The green surroundings and the woods in the small river valleys were kept to frame in groups of row houses with separate playgrounds and footpaths. This was an important idea, according to the retired architect:

> It is eight cooperative housing units and some blocks of flats, and they all are situated in flat areas. Then there are some ravine valleys. Much of this [the row houses] are placed in such a way that there is woodland in between, so we cannot see each other. So, even if it is in sum seven, eight thousand [people], the communities are the physical units.

The tight and thoughtful organization of space, coming from the modernist tradition, coupled with a kind of micro-spatial sensitivity and flexibility (not found in large-scale modernist estates), is of course not without problems or resistance. There are stories of opposition, of youngsters hanging out in kindergartens and playgrounds where trampolines are an issue of conflict, says the planner who had lived in Skjettenbyen in the 1970s. Besides, places and housing areas change, and changes are viewed differently because of social differences. One informant, a retired man, focused on the element of decay: 'There are always people who don't bother to mow their lawn, while the neighbour tries to decorate. Such things make me irritable'. Stories of neglect may be contrasted with narratives of intense commitment to the aesthetics of the place, such as those people who turn their small plots into beautiful gardens. Even if Skjettenbyen is a large-scale project represented as a scientifically planned and industrially produced place, the degree of flexibility and participation in future developments, within the limits already discussed, make room for individual expressiveness to some extent.

With time, not only do the social practices and the cultural conceptions of landscape change, but the landscape itself also changes. The green maturing of Skjettenbyen with large trees and dense vegetation cover has given the place a *patina*. There seems to be experiential evidence that the housing project has matured into a liveable place, both socially and aesthetically. Social problems related to unrest because of local youth hanging out are less visible, and the landscape has matured. This full-grown and relatively well-kept vegetation cover is a sharp contrast to the barren lands and muddy soil of the pioneer settlement of Skjettenbyen. The spirit of settlers is not only a cultural trait, but also a landscape representation, as Figure 7.2 illustrates. This feeling of being set in a non-urban landscape, with agricultural fields nearby is also part of this, expressed for example by a Pakistani man in his 40s, with children:

> […] all the surrounding fields we can see, they will disappear sooner or later. There is much to be gained from selling and making houses […] seen from my point of view, the area will be poorer […] this was attractive for us, amongst other things. There were fewer houses, more green areas and less traffic. All this in combination [...].

The Skjettenbyen project itself of course changed the landscape and social structure of the existing local community. There is a lost landscape as well with images engraved in peoples' memories. A woman in her 40s who grew up in her area shared her memories of how it was before the landscape was bulldozed and houses built:

> I think I was seven years old when they started building. But I can remember us skiing here, probably also because we were told that it was so nice here, before these buildings came […] when Skjettenbyen came, it was at once very many [people] at one place. The school was more than full, and this was in a way not planned. The surrounding area was not ready to receive that many people […] it became a gang, such juvenile problems we didn't have before.

Skjettenbyen may share the characteristics of a New Town, being placed on 'undeveloped' land, but it is more of a satellite because of the limited range of services and work places. The representation of Skjettenbyen as an isolated enclave, in spite of severely increased spatial mobility, has been kept because of its 'inward' oriented design, and because only parts of the surrounding agricultural fields have been developed into housing areas. Being a relatively large enclave does make it even more attractive to have a fully developed system of recreational grounds and landscaped surroundings.

Even if the flexibility is mainly a spectacular feature of the architecture itself, the Skjettenbyen project and its history of practices signal an aura of flexibility and adjustment that fits into a postmodern participation perspective, even while the frame of design is controlled through a strong discourse and practices. This strong planning discourse has produced a structure, which despite its relatively high density and the inherent opportunities to alter the design detail, has provided recreational spaces at different levels and for a variety of purposes.

Conclusions

The conception, design and greening of suburbia are closely connected to the fate of the green urbanism project. For practical reasons it is highly unlikely, even if desirable, for massive re-centralization of urban populations. As long as suburbs exist, their character becomes an important part of the green urban project. Arguably, most people live in suburbs, and cities will not be resilient or green if their suburbs are not. However, the urban-suburban dichotomy may be problematic, because 'suburban lifestyles' do not seem to be confined to the suburbs, but also characterize the movement back to, and the re-urbanization of, cities. Butler (2007) argues that gentrification is the saturation of cities with the lifestyles of suburbia. On the other hand, because of the increasing differentiation of suburbia, and the social filtering of certain suburbs, substantial parts of suburbia may, paradoxically, not be characterized by traditional suburban (middle and upper-middle class) lifestyles. Thus the complexity of the urban region must be taken into consideration. There is a need to break loose from dominating representations of the suburban, as well as urban.

This chapter contributes to an effort to differentiate the image of suburban greens, and to investigate how the green spatiality has constituted the idea of suburbia and the planning of modern suburbs. The design of the spaces between houses and the representation of the suburban green has been a focal point of how suburbs have been conceived and conceptualized. However, there has been little focus on the differences within suburbia, which are apparent and have been from the start of the process of modern suburbanization itself. The contrast between the privatized and excessive greens of the picturesque enclaves and the common recreational grounds of suburban high-rise estates is indicative. However, the conceptual binary of the private green versus the green commons may obscure both the socio-spatial variation in today's suburban landscapes and the differentiation in how green spatiality has been represented and implemented in suburban contexts.

The aim of this chapter has been to uncover some of these variations and their representations, focusing especially on a middle ground between the two extremes, represented by Skjettenbyen. This housing project is produced through a planning discourse, arguably to reduce urban sprawl and at the same time provide for a fine meshed green and recreational spatiality. It also represents a middle ground because, while it is initiated top-down and planned nearly scientifically on a large scale, it involves ordinary people, who despite the relatively high density of the area and embedded structural aesthetic, are given the ability to change their environment through a flexible response to existing building structures. Even if it may be considered a modern project, because of its comprehensive and detailed planning, I argue that the spatial flexibility makes it fit into a postmodern reaction against modernist top-down planning, but one that is far from the postmodern highly privatized suburbanization, or exurbanization, as described by Knox (2008).

The case of Skjettenbyen illustrates how an ideologically and discursively saturated planning project may produce environments that resonate with peoples' experiences of a place for living, both in the functional perspective of everyday practices, but also in the constituted meanings developed by its inhabitants. Research in other contexts also emphasizes the importance of parks and gardens and informal green areas significant for social encounters (Burgess, Harrison and Limb 1988). Careful studies of this interplay is needed in the search for knowledge to develop a differentiated and context-sensitive approach to green urbanism not hindered by the urban-suburban divide, and at the same time achieve a more just city that makes green spaces available for people all over the city. This is a timely ambition because green suburbanism may well become an asset, as it is in private or corporate master-planned communities catering to a consumer-oriented housing market where urban politics and planning are receding, making a stylized environmentalism part of an increasing social polarization. The public value of green urbanism may then deteriorate, leaving us with an increasingly fragmented social landscape.

Bibliography

Askew, L.E. and McGuirk, P.M. 2004. Watering the suburbs: Distinction, conformity and the suburban garden. *Australian Geographer*, 35(1), 17–37.

Brattbakk, I. and Hansen, T. 2004. Post-war large housing estates in Norway – Well-kept residential areas still stigmatised? *Journal of Housing and the Building Environment*, 19, 311–22.

Burgess, J., Harrison, C.M. and Limb, M. 1988. People, parks and the urban green: A study of popular meanings and values for open spaces in the city. *Urban Studies*, 25, 455–73.

Butler, T. 2007. Re-urbanizing London Docklands: Gentrification, Suburbanization or New Urbanism. *International Journal of Urban and Regional Research*, 31, 4, 759–81.

Duncan, J.S. and Duncan, N.G. 2004. *Landscapes of Privilege: The Politics of the Aesthetic in an American Suburb*. New York: Routledge.

Feagan, R. and Ripmeester, M. 2001. Reading private green space: Competing geographical identities at the level of the lawn. *Philosophy & Geography*, 4(1), 79–95.

Fishman, R. 1989. *Bourgeois Utopias*. New York, NY: Basic Books.

Frampton, K. 1992. *Modern Architecture: A Critical History*. Third edition. London: Thames & Hudson,.

Gans, H. 1967. *The Levittowners*. New York: Pantheon.

Gehl, I. 1973. *Bo-miljø* (Dwelling environment). SBI-rapport 71. Copenhagen: Statens byggeforskningsinstitut.

Guiton, J. 1981. *The Ideas of Le Corbusier: On Architecture and Urban Planning*. New York: George Brazillier.

Guttu, J. 2003. *'Den gode boligen': Fagfolks oppfatning av bokvalitet gjennom 50 år* ('The good dwelling': Professional views on dwelling quality during 50 years). Avhandling 11, Oslo School of Architecture, Oslo.

Hall, P. 1982. *Urban & Regional Planning*. 3rd edition. London: Unwin Hyman.

Hall, P. 2002. *Cities of Tomorrow*. Blackwell, Oxford.

Hanlon, B., Short, J.R. and Vicinio, T.J. 2010. *Cities and Suburbs: New Metropolitan Realities in the US*. Routledge, Oxford.

Harris, R. and Larkham, P. 1999. 'Suburban foundation, form and function', *Changing Suburbs: Foundation, Form, and Function*, edited by R. Harris and P. Larkham. London: E & FN Spon, 1–31.

Hayden, D. 2003. *Building Suburbia: Green Fields and Urban Growth, 1820–2000*. New York: Vintage.

Jackson, K. 1985. *Crabgrass Frontier: The Suburbanization of the United States*. New York: Oxford University Press.

Knox, P. 2008. *Metroburbia, USA*. New Brunswick, NJ: Rutgers.

Lefebvre, H. 1991. *The Production of Space*. Oxford: Blackwell.

Lorange, E. and Myhre, J.E. 1991. 'Urban Planning in Norway', in *Planning and Urban Growth in the Nordic Countries*, edited by T. Hall. London: E & FN Spon, 116–66.

Lund, N. 1973. Skjettenbyen, Norge (Skjettenbyen, Norway). *Arkitekten*, 10–11.

Newman, P.W.G. and Kenworthy, J.R. 1980. *Cities and Automobile Dependence. An International Sourcebook*. Aldershot: Gower.

Nilsen, K.A. 1998. *Nordmarkstrikken: Holmenkollbanen gjennom 100 år* (The 'Nordmarka" tram: 100 years with 'Holmenkollbanen'). Oslo: Aschehoug.

Næss, P. 2006. Accessibility, Activity Participation and Location of Activities: Exploring the Links between Residential Location and Travel Behaviour. *Urban Studies*, 43(3), 627–52.

Røe, P.G. 2009a. The construction of a suburb: Ideology, planning and identification in 'Skjettenbyen'. Paper presented at the conference *The Diverse Suburb: History, Politics and Prospects*, New York: Hofstra University, October 22–24, 2009.

Røe, P.G. 2009b. Forstaden som sted (The suburb as place), in *Mennesker og steder i samspill* (People and Places in Relation), edited by N.G. Berg, S. Jørgensen; A. Karlsen and A. Aase. Tapir Akademisk Forlag. Trondheim, 43–60.

SBI, 1971. *Tæt lav – en boligform. Eksempelsamling* (Dense-low – a Type of Dwelling. Collection of Examples). SBI-rapport 75. Statens byggeforskningsinstitut, Copenhagen.

Short, J.R. 1989. *The Humane City*. Oxford: Blackwell.

Talen, E. 1999. Sense of community and neighbourhood form: An assessment of the social doctrine of new urbanism. *Urban Studies*, 36(8), 1361–79.

Thorén, K.H. 2002. Utearealene (The outdoor areas). *Fremtid for fortiden*, 3/4.

Whitehand, J.W.R. and Carr, M.H. 1999. 'England's garden suburbs: Development and change', in *Changing Suburbs: Foundation, Form, and Function*, edited by R. Harris and P. Larkham. London: E & FN Spon, 76–90.

Chapter 8
Driving Spaces and the Dilemma of the Green City

Even Smith Wergeland

Introduction

In this chapter, I investigate the troubled relationship between the green ideology of the contemporary urban discourse and the car-based ideology of modernist urbanism.[1] In spite of its powerful body of well-crafted arguments, the green agenda fails to deal with certain aspects of the contemporary urban landscape: the unruly, disturbing and dangerous places in a city, such as driving spaces. The ongoing celebration of the green, I argue, disqualifies these spaces from being included in thoughts and theories on the urban. The chapter seeks to highlight how a set of interesting car-based planning principles from the past have been hidden from view by a one-dimensional focus on sustainability and similar contemporary manifestations of green urbanism.

I explore this issue by studying the significance of 'driving spaces' in relation to the idea of the green city, in architectural and urban history as well as contemporary manifestations of this constellation. This exploration is undertaken through a discussion on theoretical positions that have been influential with regards to the topic of roads in urban planning, for example Le Corbusier's obsession with car traffic in the 1920s and 1930s, and Alison and Peter Smithson's radical theories on urban motorways from the 1950s and 1960s. Then I relate these theories to Oslo, and present a case, a study of one specific urban road, Kirkeveien [The Church Road]. Kirkeveien is chosen because of its long history as an urban road in Oslo, because of its significant presence in the contemporary cityscape and, most importantly, because of the continuous negotiation of it as a green route.

It is not my intention to revive modernist city planning as a possible solution to current issues at stake in Oslo, but to engage critically with one of its specific legacies, the urban road system. In so doing, the chapter has a twofold purpose: to shed light on some of the often overlooked subtleties of this tradition and to problematize certain aspects of the current reception of it. A notable fault with the green discourse of the present is, I would argue, the rejection of modernist

1 In this context, the term 'car-based ideology' applies to movements within urbanism and architectural theory that propagates a positive view on urban roads and motorways by means of theoretical proposals as well as operative urban strategies.

Figure 8.1 Near the vertex of Kirkeveien at Marienlyst

Source: Photograph provided courtesy of the author, 2010.

urbanism either as a purely utopian discourse or as a general failure.[2] For example, when Stephen M. Wheeler and Timothy Beatley introduce Jane Jacobs's critique on modernist urbanism they claim that 'Jane Jacobs helped lay the groundwork for a renewed emphasis on "place-making" with her critique of the sterile, automobile-oriented urban landscapes created by much mid-twentieth century modernist architecture and urban renewal' (Wheeler and Beatley 2004: 73). Not only does this line seem to suggest that modernist planning was placeless, as if Jacobs and her followers invented the idea of place and local environments, it also reveals a quite one-dimensional conception of the legacy of the modern movement. The modernist urban landscapes are frequently described as sterile, bleak, misconceived and inhuman throughout the book. They are the consequence of inconsiderate and poor planning. Accordingly, planners promoted freeways and other automobile infrastructure without considering their sprawl-inducing impacts, authorized by the bulldozing of vibrant older urban neighbourhoods for redevelopment into bland, modernist apartment blocks, allowing the destruction of natural landscape features such as streams and wetlands, and aiding in the

2 The expression 'purely utopian' should in this context be understood as a discourse which is unable to deal with factual issues and real places.

segregation of racial and socio-economic groups through zoning and red-lining (Wheeler and Beatley 2004: 30).

Despite the well-crafted arguments that make up this body of critique I will argue nevertheless that parts of it build on unfortunate misconceptions and simplifications, which are being amplified by the green agenda of contemporary urbanism. To rephrase the issue, I also suggest that there are many problematic and contradictory aspects within the discourse of green urbanism itself. Can the city really be a landscape, as suggested in Tom Turner's analysis (Turner 1996) of the post-postmodern view on design and planning? The celebration of the green, either as on overall ideal in accordance with Turner's approach or as a guiding principle for specific functions demonstrates that green urbanism has a distinctly dogmatic side to it. When critiquing this dogmatic dimension, I am particularly concerned with the heavy reliance on 'environmental-friendly' technologies to aid the development of new urban schemes, an approach that, in my view, often appears in the guise of clever marketing campaigns rather than workable solutions to urgent problems. At a more theoretical level, I look into the widespread insistence on the positive value of concepts like 'neighbourhood scale' (Cahill 2003: 599–603, Wheeler and Beatley 2004: 30–4) and 'Integration planning' (Miller and de Roo 1999). I highlight how these concepts are productive to the continuous confirmation of green urbanism as a positive thing and, vice versa, counter-productive to a thorough way of understanding and dealing with modernist driving spaces.

The Rise of a Green Agenda

Half a century ago, roads were hailed as icons of modernity and were ever-present in glorious prophecies about the future. A typical example is Norman Bel Geddes' Futurama exhibition at the New York World Fair in 1939/1940: a set of futuristic urban layouts completely governed by urban motorways (Bel Geddes 1940). Bel Geddes envisioned a modern society motoring towards prosperity, with roads that could deliver efficiency and freedom for the individual. Today, however, roads for car traffic signify interruption and failure, especially in the case of large-scale inner city urban motorways, which tend to be camouflaged via fenders or hidden in tunnels. While cities continue to build roads they normally appear in the shape of small-scale streets with reduced speed. In Norway, this street model is commonly referred to as *strøksgater* or pedestrianized street, a road design that has been implemented in several cities over the past ten years (Naper 2005). The typical post-war urban road, however, is seen as an obstacle in the urban way of life. The future is green and it belongs to collective transport, pedestrians and sustainable design, as suggested in key literature on green urbanism (Miller and de Roo 1999, Beatley 2000, Jefferson, Rowe and Brebbia 2001, Wheeler and Beatley 2004, Jencks and Dempsey 2005).

It would be hugely incorrect to suggest this green agenda is a new thing. As others have noted, the discursive power of sustainability was well established by the beginning of the 1990s (Owens and Cowell 2002: 2–3). It is true that the first decade of the 2000s has brought about an even stronger version of this movement, fuelled by greater financial support, political influence and rhetorical firepower. Along with writers such as Peter Newman, Gert de Roo and Donald Miller, Tim Beatley has been a key person in propagating the importance of the green agenda. According to Beatley, we need to devise an entirely new urban strategy to improve the quality of urban living in contemporary cities:

> Green urbanism emphasizes the important role of cities and positive urbanism in shaping more sustainable places, communities and lifestyles. And, it implicitly emphasizes that our old approaches to urbanism – our old views of cities, towns, and communities – are incomplete and must be substantially expanded to incorporate ecology and more ecologically responsible forms of living and settlement. (Beatley 2000: 5)

While Beatley does not mention the issue of cars and roads directly in this particular publication, he does in *Resilient Cities* (Newman et al. 2009: 3–5). Written in collaboration with Peter Newman and Heather Boyer, the book focuses mainly on the urban road as a generator of continued dependence on petroleum, and hence a contributing factor to global warming. Where they do address the topic of roads in greater detail, it is to demonstrate the need to remove the road system rather than adapting it to current needs (Newman et al. 2009: 86–112). Newman, Beatley and Boyer's critique of old school urbanism is quite evidently an attack on modernist urban planning and its firm belief in the splendour of the motor car. The modernist urban discourse is understood as a negation to the idea of a positive urbanism, which they argue requires that sustainable measures must be the top priority of urban policy.

This antagonism toward modernist planning, particularly the large-scale urbanism that emerged in the post-war period, arguably stems from the seminal work of Jane Jacobs in the early 1960s (Jacobs 1961). Jacobs's main agenda was to save green areas and old neighbourhoods in American cities by protesting against huge development projects such as urban motorways and high-rise apartment blocks. Her countless battles with Robert Moses, the controversial New York urban planner, have become a symbol of grassroots activism, a heroic tale of successful resistance (Flint 2009). During the 1950s and 1960s, as Moses's planning empire started to lose ground, Jacobs was there to challenge him both directly, via organized protests and petitions against Moses's projects such as the Cross-Manhattan Expressway, and indirectly via her critique of the planning tradition he represented.

Jacobs's rage against the modernist 'machine society' is still apparent today as architects, urbanists, landscape architects and academic scholars are answering to the quest for new alternatives. Many of her ideas sit well with current concerns, especially her preference for the local, intimate and lively 'neighbourhood scale' society. Her groundwork has undoubtedly brought about necessary changes in our conception of urban spaces and how to deal with them. On the other hand, one could argue that the Jacobs's influence has helped create a very predetermined, formulaic-based idea of what a good city should be. While the emphasis on soft urbanity, for example parks, street cafes, playgrounds, low-rise dwellings and so on may be good for some functions, a city cannot prosper entirely on these ideals. Jacobs's theories remain largely inattentive to the industrial scale needed for some operations in a city.

If Jacobs represents the small-scale preservationist aspect of the green agenda, other agents find themselves on a different quest. Spurred on by predictions of future global warming and other acute environmental issues, some architects and urbanists are working on more drastic solutions, seeking to go beyond the existing urban environment and replace it with entirely new layouts. It must be remarked, however, that not all the propagators of green urbanism advocate this extreme scenario. Susan Owens and Richard Cowell, for instance, argue that even a specific catchment area like 'integration planning' is too evasive to produce concrete results: 'While the rhetoric of integration is pervasive, its precise meaning and its implications for policy practise are often unclear' (2002: 64). But still, in contemporary architecture and urbanism there has been a wave of extremely ambitious projects of exactly the kind that Owens and Cowell question.

A typical example is the 2009 *Fremtidens arkitektur er grøn!* [*The Architecture of the Future is Green!*] exhibition at the Louisiana Art Museum in Denmark; there a selection of world-famous architectural offices were invited to share some of their latest projects and proposals. The exhibition dealt explicitly with current buzzwords like 'sustainability' and 'eco-friendly' architecture. It was marked by a distinctly architectural approach, that is, on buildings and design methods, but its main focus was the application of eco-friendly principles onto the urban scale. Due to a clever presentation of the subject matter, which was promoted through 3D projections and spectacular models, the exhibition managed to convince both the general public and professionals. As one reviewer put it: '*Fremtiden Arkitektur er grøn!* is an important exhibition. By means of optimistic, inventive and diverse examples the exhibition shows that sustainability and aesthetics cannot be understood as separate concepts but as two important tools that can work interdependently in the development of a new architecture. Go to Louisiana, be inspired and dream about the architecture of the future' (Lund 2009: 1).

Unfortunately, though, the reviewers and other commentators failed to address the evident shortcomings of the exhibition. The persuasiveness of the material could not disguise the fact that many of the proposals relied on technology yet to be properly developed. Also, a significant number of the proposals on display were based on a complete refurbishment of existing urban spaces, but the immense economic challenge of having to rebuild entire cities, which would not be a very sustainable move, was hardly touched upon. In that sense, the exhibition was just as naive as the utopias of the past. The exhibition was, ironically enough, an effective reminder of the optimistic spirit of the post-war period when obsession with progress and future-oriented solutions sometimes would cloud the power of judgement and diminish critical self-reflection.

It is perhaps overly critical to expect precision and perfection at every level from a one-off event like *Fremtidens arkitektur er grøn!* But when the high level of prestige and media coverage is taken into consideration, the exhibition must be regarded as an influential force and therefore analysed accordingly. I would in fact argue that *Fremtidens arkitektur er grøn*! is symptomatic of a problematic aspect of the contemporary urban discourse, namely that certain issues are avoided and several questions left unresolved in order to establish a coherent doctrine, a formula for people, politicians and developers to believe in. The technological and economical dimension is one thing. The lack of concern for past ideas and existing structures is far more serious. For instance, the anthology *Future Forms and Design for Sustainable Cities* sets out to '... present the reader with examples of the latest research into different urban forms and the ways in which they can be designed to be more sustainable' (Jencks and Dempsey 2005: 13). Accordingly, the book contains a wide range of thorough mappings and interesting suggestions. Despite the interesting tools and convincing statistics, the anthology is weakened by its obsession with how to implement new ideas. The reader is quickly exposed to the notion that the existing urban structures have no real qualities, apart from the fact that the concept of the city is the best way to achieve a sustainable society (Jenks and Dempsey 2005: 13). The various entries in the anthology add new ingredients to the formula, but they fail to convince, either with regards to historical insight, or with their handling of the present.

While these examples represent a significant degree of unrealistic utopianism, there are also more moderate thoughts in circulation. *Resilient Cities* exhibits a similar utopian feel, but Newman et al. acknowledge that any fundamental changes in technology and planning must 'involve change in our cultures, our economies, and our lifestyles' (2009: 4). These differences aside, the 'radicals' and the more moderate thinkers do have one thing in common: they underplay the significance of the existing urban road network. When it is not underplayed or overlooked, it is normally the subject of a fierce one-way criticism, serving only as an example on how bad things can get. When transportation is dealt with, for instance in *The Sustainable Urban Development Reader* (Wheeler and Beatley 2004: 87–110), roads are viewed entirely as a thing of the past. They are the principle target for the manifestation of the language of contemporary green urbanism, which defines itself via a series of rhetorical oppositions in relation to the post-war period: from hard to soft, from alienating to friendly, from machine-like to human, from grimness to beautification, from monotony to diversity, as evidenced by Peter Newman and Jeffrey Kenworthy's pivotal text on traffic calming (1989).

This tendency is problematic for three reasons. First, the road networks have a complex history as infrastructural tools, planning elements and aesthetic objects that need to be understood properly before radical decisions are made. They are absolutely essential to how our cities appear today. Second, they still take up a significant position in contemporary infrastructural planning. We may like it or not, but society still builds new roads. In Oslo, the post-war road system prevails while new elements consistently are being added. Third, the roads constitute large structures that occupy vast urban spaces. To remove them would by no means be a simple operation even with a solid financial backbone to aid the operation. The reduction of driving spaces from four to two lanes to accommodate existing systems like light rail and tram lines could be an alternative but, as Owens and Cowell argue, planning projects that support these options have often 'failed to engage with the car as a commodity and icon, neglecting profoundly influential aspects of "car culture" that form an important of the context for transport policy' (2002: 94).

Owens and Cowell remind us that agents who subscribe to the ethos of green urbanism sometimes underestimate the symbolic, aesthetic and cultural power of the car and its various manifestations. Alternative options, however ingenious they may appear to be, will have to face the real symbolic power of the motor car to be able to break new ground. In addition, the implementation of new transport systems is further complicated by the changing nature of car culture. When reading about sustainability and eco-friendly solutions one often gets the impression that this culture died sometime in the past. We are, in fact, witnessing a rapid restructuring of the car industry through a concentration on hybrid motors, electric vehicles and other eco-overtures. One could certainly ask whether is it sustainable to remove the road system at a point when the prospect of designing environmental-friendly cars seems more attainable than ever.[3] At the present, the idea of a green car is not necessarily more utopian that the key ideas that constitute green urbanism. It is an element of some concern that green urbanism insistently relies on some technological advances but remains ignorant of others. The bottom line is that as long as there are roads for motorization about, the topic needs to be addressed in a discussion of the green city.

The Road and the Green City

To understand the relationship between the road and the idea of the green city, it is necessary to make a historical journey back to when this connection became a real force in urban theory. One should think that a natural starting point would be Ebenezer Howard's seminal publication *Garden Cities of To-Morrow* (1902), a book that influenced the planning of Oslo in the 1920s (Aspen 2003). But although the garden city model does contain a ring road prototype displayed as six

3 On the one hand, there are agents like http://www.cleangreencars.co.uk/ and http://www.hybridcar.com/ that work to improve existing solutions like hybrid fuel and electricity. In addition, there is an abundance of experimental projects, such as General Motor's new urban mobility vehicle, a futuristic pod car, and the Magnetic Air Car. See http://www.wired.com/autopia/2010/03/general-motors-en-v/ and http://magneticaircars.com/index.html (both accessed: 30 October 2011).

radial boulevards, Howard's main concern is how to restrict the impact of traffic. According to him, the road is not a positive resource for the urban landscape except as an efficient tool for transportation, and only if handled with proper caution. Caution is not an option for Le Corbusier 20–30 years later as he tries to reinvent the green territories of the modern city by the invention of the City of Tomorrow, in which huge motorways handle heavy loads of traffic and there is a distinct separation between parks and driving spaces (1929). This is a well-known and much criticized urban layout. If we go to one of his lesser known publications, we find that Le Corbusier did not consider roads and parks solely as separate units. In *The Four Routes* he proposes a modernist integration of roads and parks when he argues that 'a road is not just a measurable entity; a road is a plastic achievement in the bosom of nature. Geometry and nature combined have often evolved beauty, something that moved us' (1947: 31). This could be brushed aside as just another unrealistic folly drawn from Le Corbusier's obsession with modern speed culture.

However, some scholars have claimed that this perfect symbiosis has actually been achieved. Sigfried Giedion, whose viewpoint differs greatly from Le Corbusier's in many respects, points to the New York area parkways designed by Robert Moses as a successful example of a road system. Giedion's appraisal of the parkway is rooted in a specific fondness for the combination of roads and landscape, an amalgamation of motorization and nature. He claims that the parkway '... humanized the highway by carefully following and utilizing the terrain, rising and falling with the contours of the earth, merging completely into the landscape' (1967: 825).

In his praise of elegant implementations of roads into nature, Giedion remains woefully ignorant of problems like pollution and traffic noise. And while it may be possible, as Giedion believes, to create aesthetically pleasing driving spaces in nature, a road set in a strictly urban setting is a different case. This problem is thoroughly dealt with in post-war urbanism, not least in the work of Alison and Peter Smithson. The Smithsons' keen fascination with the concept of mobility is the basis of their search for a more flexible urban system, put on display with the invention of the cluster city in the late 1950s. By means of a comprehensive network of urban motorways, the cluster city is devised to offer its inhabitants an efficient system of interconnectivity and give them access to a range of different facilities. The Smithsons claim that the motorway is not only a liberator of movement, but also a liberator of diverse land use: a route into nature and an instrument for securing green spaces in the city.

The Smithsons also point to the structural and aesthetic similarities between roads and nature: 'The most important thing about roads is that they are big, and have the same power as any topographical feature, such as a hill or a river; to create geographical, and in consequence social, divisions' (1967: 50). Just like a river or a beautiful landscape route, urban roads bring an aura of quietude and tranquillity into the metropolis. They illustrate this idea with metaphorical imagery: a mixed-media collage of written text and drawings displaying abstract patterns of flow in which the inter-connective qualities of motorways are equalled with rivers, harbours and canals (Ibid.: 64). The visual potency of the Smithsons' motorway poetry fails, just as Giedion's before them, to address issues of noise and pollution.

The visionary shrewdness of the Smithsons' juxtaposition of motorways and natural landscape is unrivalled in its radicalism. But the idea of the road as liberator of green spaces is also found in the other outputs from the post-war period, for instance the work of the British Townscape movement. Townscape members were inspired by older traditions in British landscape theory, such as Uvedale Price's writings on the picturesque in the late 18th century (1796), and for this reason they are often viewed as polar opposites to the Smithsons. But Townscape does not merely represent traditionalism. This tag has often been assigned to the movement because the most well-known Townscape publication, *The Concise Townscape*, coincidentally was released during a time when a number of books were critiquing modern architecture. However, as the architectural historians John Macarthur and Mathew Aichison point out, the earlier versions of the book were explicitly modernist (2010: 14).

In the 1950s, Townscape was firmly placed within a modern idiom, a position that culminated in 1961 with the publication of an experimental project called *Motopia* (Jellicoe 1961). This book was written by the influential landscape architect G.A. Jellicoe in collaboration with Gordon Cullen, the architect who wrote and illustrated the various versions of *The Concise Townscape*. As two of the most prominent members of Townscape, Jellicoe and Cullen's joint output highlights the movement's more progressive side. *Motopia* is an urban project founded on a desire for garden design and road construction in equal measures. The master plan consists of a rectangular grid system quite similar to Le Corbusier's plan for the *Radiant City* from 1935. A structure like this was not really avant-garde in the early 1960s, but *Motopia* does have one feature of a highly radical nature. The whole system of main roads and motorways is placed on top of the roofs of the city's dwellings, leaving the ground level completely free of traffic and open for spacious green areas and other attractions – an ideal 'combination of the biological and the mechanical' as Jellicoe himself described it (Jellicoe 1961: 11).

Even though none of the Townscape or Smithson projects were carried out in the real world, at least not in their initial forms, they demonstrate that green initiatives and ideas were distinctly present in the modernist urban discourse and that these ideas were mediated through a range of different proposals and theoretical positions. And moreover, as Susan Robertson accounts for (Robertson 2007: 74–91) in her analysis of the Westway, an elevated motorway in London, this particular late-modernist idiom was realized through a range of projects that changed the spatial and perceptual experience of moving through modern cities.

But a wave of criticism had already surfaced long before the Westway, which was completed in 1970, most notably through the writings of Jacobs and Lewis Mumford (1961, 1964). In *The Highway and the City* Mumford describes how the motor car has devoured the green spaces of the modern city, causing it to become a barren wasteland of roads and parking lots (1964: 235–6). Opinions like this sparked a massive campaign against the negative impacts caused by heavy motorization in urban areas, inspiring a new generation of scholars to protest against cars and their corollary, the urban motorway (Newman and Kenworthy 1989, Kay 1998).

The Driving Spaces of Oslo

Many aspects of this international antagonism against the modernist legacy apply to Oslo. Even during the most optimistic decade with regards to urban roads, the critical voices of the 1960s emerged, especially in the architectural magazines, *Byggekunst* and *Arkitektnytt*. Although many architects shared the engineers and the public planners' strong belief in urban road building, individuals like Per Cappelen and Robert C. Esdaile were sceptical about the consequences of a road-based city plan for Oslo. In 1964 Esdaile posed the question: 'The car – friend or foe'? Esdaile's rhetorical question was a counteraction against the new public plans for Oslo and a retreat on behalf of the architects, who previously had been advocating the need for a large-scale urban road network in Oslo. There were only a few architects who actually worked in the municipality's planning department – one of them, Nils Haugstveit, was a key member in the team behind the 1965 Transport Analysis for Greater Oslo (Oslo byplankontor 1965) – but the road and its function in urban planning was a hugely important topic to the majority of Oslo-based architects of that period. In 1957 the Norwegian architect John Engh wrote this eulogy to the potential of the car:

> The car satisfies our deeply rooted thirst for liberty, freedom, wanderlust – our desire for contact with other people. However, the car is not an irrational toy as our authorities and a segment of the population seem to believe. In Western culture the motorcar is a necessary tool for progress. It offers greater radius of action for people and increases productivity in all industries that are dependent on rapid transport. Also, the car has a significant advantage compared to other means of transportation because it is individual. (Engh 1957: 73)

Although Engh acknowledges the dangers of unleashing the car on the urban environment, his statement encapsulates the positive view that many Norwegian architects had in the early post-war years. The optimistic spirit was further galvanized by certain international impulses at play. In 1960 *Byggekunst* released a special section highlighting theoretical perspectives and practical approaches to the planning of Oslo (Larsen 1960: 197–224).

The opening of the section is written by Peter Smithson whose lecture for the Oslo Architectural Association (OAF or Oslo *Arkitektforening*) the previous year clearly had earned him some recognition. His entry deals with the fundamental principles of what he calls 'fragmentary Utopias'. 'We must plan as we must live', he writes, 'towards a dream – even if that dream has built-in unrealisability' (Smithson 1960: 1). Behind these rather quirky words lies an infinite belief in urban planning that relates to the radical impact of motorization: 'The motor car, combined with a high level of urbanization, have made the existing archetypal solutions, e.g. "shopping street", "town hall", "apartment house", and so on, obsolete' (Ibid.). Smithson's message is that planners who wish to operate successfully in a situation like this must embrace radical means and methods. He concludes the entry by proposing an 'open aesthetic', a flexible mode of thought that allows for a close understanding of sites and contexts without succumbing to the shortcomings of the English Picturesque or Camillo Sitte, on whom he frowns in the article (Smithson 1960: 2).

While Smithson dwelled on abstract concepts and theories, other foreign contributors focused on more concrete matters in the development of Oslo. A recurring guest on the Norwegian architectural scene was the Greek architect and urban planner Constantinos Apostolos Doxiadis. Doxiadis's first of many visits to Norway was in 1960, when he was invited to give a lecture at the Oslo Architectural Association. In this lecture he presented a list of principles for the city of the future, Dynapolis; a concept that aspired to deal with 'all the problems of the past, present and future' (1960a: 1). His concluding remark was that 'Every problem that occurs can be solved, not by adopting the ideal scheme described in this lecture, but by using it as a source of inspiration for our future plans' (Ibid.: 77).

Immediately after the lecture, Doxiadis was encouraged to make a sketch for a master plan for Oslo. His proposal was published in *Byggekunst*, where he shared his thought on the matter and presented two illustrations to complement his arguments (Ibid.: 25–8). The proposal contains his trademark urban grid, characterized by a circular classification of zones and a huge circulation system with one main road. Doxiadis himself was quite humble with regards to the proposal because he was aware of his own lack of detailed insight into local matters. His Norwegian colleagues, however, were thrilled by his effort:

> With the ongoing discussion on the city plan for Oslo in mind, one must say that dr. Doxiadis' lecture in the OAF was a thought-provoking experience for many listeners. He presented his viewpoints in a very precise and accurate manner and accounted for the main principles behind the concept of Dynapolis, the dynamic city, the ideal plan for the future. (Cappelen 1960: 67)

Doxiadis and Smithson's idea of a dynamic, flexible city was obviously regarded as a fresh perspective in the ongoing debate. And the fact that cars and roads played a decisive role meant some elements from their work could be appropriated almost directly into the ongoing planning of Oslo. One of the big issues at stake in Oslo was how to combine a dynamic infrastructural system with the existing green structures and other urban features. Even the 1965 Transport Analysis, a municipal document containing the most extensive urban motorway proposal in the history of urban planning in Norway, tried to take this matter into consideration: 'As far as it is possible to predict the prospective growth of traffic and all the functions of the city, the actual purpose of the transport analysis it to prevent unfortunate effects of the general plan and to encourage good urban design' (Oslo byplankontor 1965: 16). In the end, though, the final version of the document did suggest several drastic interventions in the city's parks, most notably Tøyenparken and Slottsparken. The Transport Analysis looked more like a fulfilment of Mumford's dystopian vision of the modern city as a barren wasteland, than a healthy management of green resources. The document was approved by the Oslo City Council in April 1967. But due to financial limitations and a gradual increase in public protests against the urban motorway, the council u-turned their decision in 1973 when they opted for the much more moderate and cautious *Gatebruksplanen* for Oslo [The Plan for street use in Oslo]. In only six years, focus had shifted entirely, from motorways to streets (Hansen and Guttu 1998: 58).

Some years later, in 1979, Christian Norberg-Schulz (CNS) wrote an article called '*Myten om den grønne by*' ['The Myth of the green city'], in which he reflects upon the legacy of Norwegian post-war planning. In this article he questions the very concept of the green city: 'Not long ago the "green city" was everybody's ideal. Today it is hard to understand why' (1979: 20). His attack on the myth of the green city must not be misconceived as a renunciation of green spaces as such but as a rejection of the utopias that failed to generate such spaces and, in this particular case, failed to provide the people of Oslo with the quality of life so often proclaimed in visionary proposals from the 1950s and 1960s. CNS laments the consequences of post-war planning ideals, and he urges the upcoming generation of architects and planners to engage with the existing, physical environment rather than eloping into 'remote abstractions', as he puts it (Ibid.: 21).

CNS's negative verdict is closely connected to the numerous road and motorway visions that came about in the post-war period. While some of these remain unfulfilled, like the six-lane urban motorway, others made it all the way from drawing-board to reality. These have been at the receiving end of a lot of criticism over the last 30–40 years. Despite all the overall negative reception, there is at least one large-scale road in Oslo which is frequently praised and rewarded with positive remarks: *Kirkeveien*, which was presented as a representation of a successful integration of road and landscape, a 'green route' in the city.

Kirkeveien: Facilitating the Urban Green?

Kirkeveien has a long and interesting history. For well over hundred years it has gone through various stages of transformation, physically, culturally, socially and aesthetically. It is not easy to trace the exact time of its origin. The official name dates from 1855 but the road existed even before that, as a small country road between Skarpsno and Ullevål. It was formalized as a main road in 1890, when the Norwegian engineer Albert Fenger-Krogh published a plan that designated Kirkeveien as the main artery in a ring road structure stretching from Frognerplass all the way to Tøyen. Not only did Fenger-Krogh draw up the ground structure of what is known today as Ring 2, he also initiated the green label by proposing it as

Figure 8.2 Green spaces unfold in the centre strip of Kirkeveien

Source: Photograph provided courtesy of the author, 2010.

a 'Parisian boulevard' for Oslo, with trees and systematically arranged vegetation (Lexau 1993: 40).

Despite this early proposal, Kirkeveien did not become a proper, purpose-built road for modern traffic until the late 1920s. In the mid-1920s it still had the character of a self-grown route, muddy and casual looking. But in 1929 it was given a major part in the *Generalplanen* [master plan] for Oslo. Harald Hals, the head of city planning in Oslo at the time, conceived the first comprehensive road plan for the city, a plan in which Kirkeveien was transformed into the 'Store Ringgate' [the Greater Ring Street] scheme. Due to Kirkeveien's prominent position in this structure, Store Ringgate would soon be known as 'Kirkeveiringen' [The Church Ring Road]. As a result of this plan, Kirkeveien finally came into being as a modern piece of road construction. Construction went on for 20 years, slowly fulfilling the 1929 version of the road. On its completion in 1950 it became the first stretch of the Oslo ring road system in full operation.

An essential part of the Kirkeveien proposal was to link the road with the green wedges of the city: Frognerparken, Akerselva, Dælenenga-Grefsen, Tøyen and Lodalen. The idea behind this link-up was to bring nature into the city and to enrich its local environment with green elements: a landscaped thoroughfare resembling a Robert Moses parkway, but in reverse. Instead of bringing commuters out in nature as Moses had hoped to achieve, Kirkeveien was supposed to generate and

Figure 8.3 A traffic jam during rush hour near the interchange at Majorstuen

Source: Photograph provided courtesy of the author 2010.

planning department's inclination for road plans of almost epic proportions, meant that more cars were coming into the city. As a natural link between the bigger roads, Kirkeveien was deeply affected by this. The result of this development is that Kirkeveien today has an average rate of 20,000 vehicles passing through every 24 hours (Oslo commune 2009: 7).

Interestingly enough, the idea of Kirkeveien as a green route has endured in spite of the considerable traffic strain. In 2006 it was nominated for Vakre Vegers pris [The Norwegian Road Directorate's Annual Award for Beautiful Roads], and achieved honourable mention in the final round of the competition. The jury was particularly pleased with how Kirkeveien 'mediates the transition from Frognerparken to the city in such a beautiful manner that the park and the street almost become one entity' (Statens vegdirektorat 2006: 89).

Once again, Kirkeveien is confirmed as an ideal integration of park and street. But before one confirms Kirkeveien as a real-life achievement of Le Corbusier's vision in *The Four Routes*, one has to take into account that the appraisal only

4 In 1948 there were 24,000 motor vehicles in Oslo. In 1975 the number was 150,000. See Knutsen, Sverre and Knut Boge, *Norsk vegpolitikk etter 1960 – stykkevis og delt?*, Oslo: Cappelen, 2005, p. 129.

Figure 8.4 Alignments of trees along Kirkeveien near Frogner Stadium

Source: Photograph provided courtesy of the author, 2010.

applies to a limited part of Kirkeveien, namely the newly refurbished stretch from Frognerplass to Frogner Stadion.

However, Kirkeveien is also maintained as a green route by various public institutions. In June 2009 the National Road Association [Statens vegvesen] put out a discussion document where new improvements of Ring 2 were outlined. One of the main goals of this plan was to '... emphasize the architectural form of Kirkeveien and to reinforce its green parkway character' (Rambøll Norge/Statens vegvesen 2009: 5). In the same document the Oslo municipality's Department of Preservation and Cultural Heritage [Byantikvaren] stressed that the original character of the road must be conserved and reinforced (Ibid.: 10). These projects, comments and statements were, in short, quite universal in their positive reception of Kirkeveien, so perhaps we should agree with Giedion's notion that roads can be 'humanized' and merged with the surrounding landscape when given the appropriate design?

There can never be a simple and straightforward answer to this question. On the one hand, Kirkeveien does represent an interesting example of modernist planning adapted to the contemporary urban environment. It echoes Le Corbusier's theory about the road as a plastic achievement; it has the interconnectivity as suggested in Smithson's cluster city; and it certainly lives up to the Townscape formula of the biological and the mechanical combined. But on the other hand, one cannot escape the fact that Kirkeveien remains a highly disturbing urban space in Oslo. During rush hour it is difficult to take notice of the pastoral aspects, and as a pedestrian you hardly pause to enjoy the green aesthetics because of the importunate traffic noise. It is of no comfort to the car drivers, either, even if they are in a position where they have time to dwell on the visual splendour of well-designed traffic islands and graceful alignments of trees. The aforementioned public document tries to deal with the dilemma of driving spaces and the green city by maintaining the visual character of the road and its green legacy, but fails to solve issues like traffic noise, unhealthy fumes and accidents – all sources of continuous protests from people who live in the area.

Conclusion

This ambiguity leaves Kirkeveien in a position that provides more questions than definite answers. The major paradox of Kirkeveien is that it simultaneously generates and prevents green qualities. More than anything else it points to the potential contradiction in postmodern green city coexistence theory in the sense that it forces us to reconsider the whole concept of the green city. The road is not a provider of a healthy urban living, as Le Corbusier envisioned it to be. But the question is: are current theories about biodiversity, eco-architecture and green urbanism capable of delivering a truly sustainable platform for the future; an alternative to the problems one encounters in a cityscape dominated by roads?

An inherited problem with some of these theories, as I have discussed above, is the heavy reliance on Jacobs's ideas and her emphasis on the 'neighbourhood scale'. This particular approach to urban planning is not compatible with all aspects of a city and its functions. Not every urban space can be a small-scale neighbourhood with parks and soft features. To make a city efficient in terms of sustainable operations like waste management, public transport and energy circulation, one needs to produce spaces that cannot be based on a human scale to function properly. It is enough to take a look at current criteria for eco-friendly houses to back this claim – the amount of special features needed to make dwellings energy efficient reduces the space left for the actual inhabitants of these houses. A city is not a house, but there is reason to believe that the impacts we are already noticing at a micro level will apply to the future refurbishing of our cities suggested by many propagators of green urbanism. This time around it may not be cars but other technological devices that will conquer the urban spaces. The quest of reclaiming the city for pedestrians and people, as advocated by Newman, Beatley and others, looks good on paper but it will have to face some of the same difficulties even should the car be abolished.

There is a social dimension to this problem as well. Green urbanism advocates biodiversity but due to certain spatial and cultural ideas the movement is sometimes rather narrow-minded when it comes to social diversity. Again, the problem is the preference for urban spaces with a selection of predefined qualities. The Jacobs tradition, for instance, represents a typical family perspective, in which issues like safety, togetherness, tradition and quietude are the major priorities. These values are often at the forefront of what is being defined as a good life in texts on green urbanism and, in more practical terms, inquiries about life quality in contemporary cities.[5] There are, however, some social groups, most notably youth and younger adults, in which many individuals do not necessarily subscribe to these qualities at all. A pleasant, quiet, café-dominated urban space, combined with the general safety concerns that mark the governance of contemporary cities, will often appear unattractive, exclusionary and pointless for people that fit into these social groups (Malone 2002: 161–4, Massey 1998: 121–9). Even the colour green, which oftentimes is taken for granted as a good thing, can be boring or even repulsive to some.

How does the latter observation relate to driving spaces? Ever since its induction into the modern urban fabric, the road has represented precisely some of the danger, roughness and speed that many teenagers and youngsters thrive upon. To underestimate car culture and its many subcultures will, as Owens and Cowell warn, be an act of disregard for the people who cherish and embrace it. *Sinsenkrysset*, a major interchange in Oslo, is for example often praised because it embodies these qualities; the noise, speed and relentless movement of the city.[6] Similar things could be included in a description of *Kirkeveien*. But this ruthless version of urbanity does not have a place in green urbanism which, as a consequence, fails to mirror the greater social diversity of a city. The integration of the green is a hollow and vacant project without a proper integration of all social groups.

In parallel with the social issues comes the question of the city as a landscape, to return to Tom Turner's framework. As much as the green agenda has its limitation when it comes to certain social groups, it is also limited, at least in some cases, by an insistence on bringing rural qualities into the urban, much in line with Ebenezer Howard's groundwork. But the city can never be like the countryside. What my analysis of Kirkeveien points to more than anything else is that the urban and the green are two incompatible concepts no matter how hard we try to redefine them. It depends, of course, on how you understand the concept of the green. As an over-arching principle for urban planning it seems futile, even with subtle strategies like 'integration planning' at hand. And when thoroughly redefined, as in the projects

5 The 2010 Stavanger Life Quality Survey [Levekår i Stavanger] is a representative example of this trend. The highly rated areas (Våland and Stokka) consist of upper-middle class family values with green spaces whereas the lowly rated areas (Badedammen and Rosenli) consist of young adult populations in dense areas with limited amounts of green spaces. See Stavanger Kommune, *Levekår i Stavanger – geografisk fordeling rapport nr. 4*, Stavanger: Stavanger commune, 2010.

6 This is how Sinsenkrysset was described when it was selected as an urban attraction in a Norwegian TV programme from 2010. See *Norsk attraksjon 6 av 7*, 3:22–3. http://www1.nrk.no/nett-tv/klipp/560503 (accessed: 30 October 2011).

exhibited at Louisiana Art Museum, it remains a big question whether it will actually provide workable solutions when visions are put into practice.

To conclude, my critique of green urbanism is first and foremost aimed at the lack of regard for certain lessons from the present reality as well as from history. We can improve things, but the urban landscape remains a case of endless contradictions and contrasting needs. Kirkeveien serves as an important reminder of this. In comparison with most of the other urban roads in Oslo, it stands out like driving space of great historical significance, a cultural monument representing many layers of aesthetic sensitivity. Through decades of moulding, reconstruction and adaptation, it has been invested with many serious efforts to handle real-life issues for the commuters of Oslo and for the people who live and reside nearby it. Therefore, it cannot simply be discarded as a remote abstraction, to reintroduce CNS's tag. On the contrary, the case of Kirkeveien encourages us to deal with the realities of what Peter Wollen and Joe Kerr call 'the new Autopia', in which the priority is '... to learn how to accommodate the car in our existing environment rather than to sacrifice those environments in pursuit of an outmoded ideal' (2002: 23). Only then can we avoid the risk of venturing into a green escapism that threatens to annihilate all other ideas about urban planning.

Bibliography

Aitchison, M. (ed.). 2010. *Visual Planning and the Picturesque – Nicolaus Pevsner*. Los Angeles, CA: Getty Publications.

Aspen, J. 2003. *Byplanlegging som representasjon: En analyse av Harald Hals' generalplan for Oslo av 1929*. Oslo: The Oslo School of Architecture and Design.

Ballard, J.G. 2008. *Concrete Island*. London: Harper Perennial.

Beatley, T. 2000. *Green Urbanism: Learning from European Cities*. Washington DC: Island Press.

Bel Geddes, N. 1940. *Magic Motorways*. New York: Random House.

Bus, A. 1999. 'Ecological renewal of post-war neighbourhoods: Problems and solutions', in *Integrating City Planning and Environmental Improvement*, edited by D. Miller and G. de Roo. Aldershot: Ashgate, 203–21.

Cahill, J.F. 2003. 'Neighbourhood-scale diversity, composition and root crowding do not alter competition during drought in a native grassland', *Ecology Letters*, 6 (7), 599–603.

Cappelen, P. 1960. Dynapolis – fremtidens by. *Arkitektnytt* (5), 67.

Cullen, G. 1971. *The Concise Townscape*. Oxford: Butterworth-Heinemann.

Doxiadis, C. 1960a. *Dynapolis – byen for fremtiden*. Lecture at the Oslo Architectural Association, Oslo, March 3rd 1960.

Doxiadis, C. 1960b. Et nytt diskusjonsgrunnlag for Oslo byplan. *Byggekunst*, 1, 25–8.

Ellefsen, K.O. 1992. 'Planlegging for det nye mennesket', in *Fortidminneforeningens Årbok*. Oslo: Foreningen, 81–95.

Engh, J. 1957. Provisorium eller langtidsplan?. *Byggekunst*, 3, 73–4.

Esdaile, R.C. 1964. Bilen, venn eller fiende?. *Arkitektnytt*, 8, 144–5.

Flint, A. 2009. *Wrestling with Moses: How Jane Jacobs Took On New York's Master Builder and Transformed the American City*. New York: Random House.

Gehl, J. 2004. 'Outdoor Space and Outdoor Activities', in *The Sustainable Urban Development Reader*, edited by S. Wheeler and T. Beatley. London/New York: Routledge, 81–5.

General Motor's new urban mobility vehicle, a futuristic pod car [Online]. Available at http://www.wired.com/autopia/2010/03/general-motors-en-v/ [accessed: 30 October 2011].

Giedion, S. 1967. *Space, Time and Architecture*. Cambridge, MA: Harvard University Press.

Hansen, T. and Guttu, J. 1998. *Fra storskalabygging til frislepp – beretning om Oslo kommunes boligpolitikk 1960–1989*. Oslo: Byggforsk.

Howard, E. 1965. *Garden Cities of Tomorrow*. Cambridge, MA: MIT Press.

Jacobs, J. 1961. *The Death and Life of Great American Cities*. New York: Random House.

Jefferson, C., Rowe, J. and Brebbia, C. 2001. *The Sustainable Street*. Southampton, Boston, MA: WIT Press.

Jellicoe, G.A. 1961. *Motopia*. London: Studio Books/Longacre Press Ltd.

Jenks, M. and Dempsey, N. (eds) 2005. *Future Forms and Design for Sustainable Cities*. Oxford: Architectural Press.

Kay, J.H. 1998. *Asphalt Nation: How The Automobile Took Over America and How We Can Take It Back*. Berkeley, CA: California University Press.

Knutsen, S. and Boge, K. 2005. *Norsk vegpolitikk etter 1960 – stykkevis og delt?*. Oslo: Cappelen.

Langdalen, E. 1962. Doxiadis i Trondheim. *Arkitektnytt*, 3, 35.

Larsen, H-K. (ed.) 1960. *Byggekunst* (8), 197–224.

Lexau, S.S. 1993. *Harald Hals' Generalplanforslag fra 1929*. Bergen: University of Bergen.

Le Corbusier 1929. *The City of Tomorrow and its Planning*. London: The Architectural Press.

Le Corbusier 1947. *The Four Routes*. London: Dennis Dobson Ltd.

Lund, A.A. 2009. Grønne drømme, *Kunsten.nu* (online 10 July). Available at: http://www.kunsten.nu/artikler/artikel.php?louisiana+fremtidens+arkitektur+klima [accessed: 10 January 2011].

Lillebye, E. 2001. 'The architectural significance of the street as a functional and social arena', in *The Sustainable Street*, edited by C. Jefferson, J. Rowe and C. Brebbia. Southampton; Boston, MA: WIT Press, 3–44.

Magnetic Air Car [Online]. Available at: http://magneticaircars.com/index.html [accessed: 30 October 2011].

Malone, K. 2002. Street life: Youth culture and competing uses of public space. *Environment & Urbanization*, 14(2), 157–68.

Massey, D. 1998. 'The spatial construction of youth cultures', in *Cool Places: Geographies of Youth Cultures*, edited by T. Skelton and G. Valentine. London: Routledge, 121–9.

Miller, D. and de Roo, G. (eds) 1999. *Integrating City Planning and Environmental Improvement*. Aldershot: Ashgate.

Mumford, L. 1964. *The Highway and the City*. New York: Mentor Books.

Naper, H.G. 2005. *Støksgater med trikk.* Oslo: Oslo sporveier.

Newman, P. and Kenworthy, J.R. 1989. *Cities and Automobile Dependence: A Sourcebook.* Aldershot: Gower.

Newman, P. and Kenworthy, J. 2004. 'Traffic calming', in *The Sustainable Urban Development Reader*, edited by S.M. Wheeler and T. Beatley. London/New York: Routledge, 97–103.

Newman, P., Beatley, T. and Boyer, H. 2009. *Resilient Cities: Responding to Peak Oil and Climate Change*. Washington DC: Island Press.

Norberg-Schulz, C. 1979. Myten om den grønne by. *Byggekunst*, 1, 20–1.

Norsk attraksjon 6 av 7, 3:22-23 [Online]. Available at: http://www1.nrk.no/nett-tv/klipp/560503 [accessed: 30 October 2011].

Oslo byplankontor 1965. *Transportanalysen for Oslo-området*. Oslo: Oslo byplankontor/Kirstes boktrykkeri.

Owens, S. and Cowell, R. 2002. *Land and Limits: Interpreting Sustainability in the Planning Process*. London/New York: Routledge.

Price, U.S. 1796. *An Essay on the Picturesque, as Compared with the Sublime and Beautiful*. London: printed for J. Robson.

Rambøll Norge AS for Statens vegvesen Region øst 2009. *Vedlegg 1: Forslagsstillers planbeskrivelse: Ring 2, strekningen Blindernveien – Vogs gate. Forslag til reguleringsplan*. Oslo: Oslo kommune.

Robertson, S. 2007. Visions of urban mobility. *Cultural Geographies*, 14, 74–91.

Smithson, A. and Smithson, P. 1976. *Urban Structuring – Studies of Alison & Peter Smithson*. London: Studio Vista.

Smithson, P. 1960. Education for Town Building. *Byggekunst Appendix*, 8, 1–2.

Statens Vegvesen 1996. *Vegvalg – nasjonal verneplan*. Oslo: Statens vegvesen/Vegdirektoratet.

Turner, T. 1996. *City as Landscape: A Post-postmodern View of Design and Planning*. London: E & FN Spon.

Stavanger Kommune 2010. *Levekår i Stavanger – geografisk fordeling rapport nr. 4*. Stavanger: Stavanger kommune.

Wheeler, S.M. and Beatley, B. (eds) 2004. *The Sustainable Urban Development Reader*. London/New York: Routledge.

Wollen, P. and Kerr, J. 2002. *Autopia – Cars and Culture*. London: Reaktion Books.

Chapter 9
Design with Nature: The Gardener's View

Per Hedfors and Clas Florgård

Introduction

On the lower-ground floor of a multi-storey car park in Uppsala, Sweden, stands a bronze statue. It was placed there to honour the so-called happy gardener, Nestor (Figure 9.1). Nestor was gardener to Queen Victoria of the United Kingdom. The placing of Nestor in the form of a statue in the garage works here as a metaphor for a neglected view – that of developing and maintaining the living city in a living landscape. This view represents an ideology that we interpret from the concept of 'design with nature'. The expression derives from the title of a book by the landscape architect Ian McHarg, published in 1969. The gardener in the garage is transformed into a malicious caricature – a negative metaphor – of a sad deportation to the underworld and the exhaust fumes, a deportation of knowledge and competence regarding the natural processes and of everything that sustains and supports life and vitality. The formerly happy gardener Nestor has been transfigured into a bronze statue and forgotten symbol banished from green parks into exile in a dark, underground car park. A sign next to the statue states that the 'happy' gardener is '… a man with a lust for life … who now has … sat down on the bench … As our parking attendant, he watches over the people and the cars. He sees life roll by'. Interpreted in this way the arrangement is humiliating to gardeners' self-image as they are banished to a 'dystopia'. The metaphor also indicates that the important knowledge of the living landscape, vital to urban development, has been exiled far below the corridors of power. The term design with nature denotes the knowledge and the attitude that has to be brought to the surface.

Our point is that the gardener deserves to be brought into the daylight where he can oversee the whole landscape, in accordance with the European Landscape Convention, which stipulates that cities and other developed areas are included in the landscape (CoE 2000). But our understanding of the gardener is more than a metaphor of banishment that ought to be reversed. The gardener in the garage stands for a practice of landscape maintenance and care and for a perspective on landscape – now slumbering in the depths of the garage – that seeks to achieve resonance between the dwellings of human beings and their surroundings. We see this as an alternative to the current approach: resonance is essential to the future not only of landscape architecture as a discipline, but of urban planning in general. In this chapter we aim to throw light upon such an alternative approach. Questions we pose are: what constitutes a practical approach within city development where

Figure 9.1 The formerly happy gardener who has been used to managing plants, soil, water, air and sunlight, is banished underground where he nonetheless manages to act as a counterweight to the unattractive environment

Source: Photograph by P. Hedfors, 2010.

rampant densification and unchecked spreading of development may be avoided? What do the planning and configuration processes that take into account the potential of the existing landscape and its innate life force look like?

The Green City

One context where a practical approach towards the existing values of the landscape is beneficial is the green city. The concept of the green city includes the preservation of green areas, but above all it aims at sustainable urban development, which, according to our perspective, means the management of living cities within living landscapes.

The concept living landscape can be seen as the blanket term for the emergent properties of a geographic framework within which all processes are contained, human activities included. The city is seen as a dense agglomeration within this wider landscape. The physical structure, the chemical flows and the biological processes of the city are significant when defining the living landscape, but so too is the involvement of inhabitants and other actors in the management of green sites. We use the concept of gardening in the broad, ideological sense where it permeates all activities of green urbanism.

The concept of the green city also comprises aspects of the biosphere which could be categorized as 'blue' (clean water) and 'brown' (management of pollution, contamination, sewage and waste). By developing knowledge derived from the natural sciences, in order to manage soil, water-flow and vegetation, and combining it with perspectives drawn from social science and the humanities, environment can be understood as a 'larger landscape' which lays the groundwork for a sustainable urban and rural development. A sustainable development is characterized by awareness and consideration of the natural processes on a site, and is based on a process of assessment and reassessment which allows future generations to alter designs and adjust practices (WCED 1987, Berg 1993).

The green city concept, and the management of the urban landscape, raises questions about the values of green environments in cities, and how these values can be articulated and developed. We look upon design with nature as a specific mode or culture of knowledge to which the gardener contributes. Landscape management applies two forms of knowledge in this respect: factual and procedural (Lang 1987, Murphy 2005). Landscape practitioners apply these two forms of knowledge, partly as scientifically grounded facts and partly in the form of the crucial competence (or technical prowess) needed to apply these facts in practice. Together they constitute a knowledge in action[1] (Ahrland 2005) which takes shape in the engineering of biotopes; by biotopes we mean the domain of plants and animals, in this case with humans included. This knowledge deserves a prominent position in the movement towards The Green City.

Design with Nature

Green urbanism has a long tradition as an environmental discourse that brings together ecological and urban perspectives. Forty years ago, the landscape architect, Ian McHarg, published his widely read book, *Design with Nature* (1969). He had a broad perspective of landscape management, integrating keystones of knowledge originally formulated by Frederick Law Olmsted, such as agriculture and forestry, technology, sociology, knowledge of nature, art and design (Roper 1973, Hall 1995, Rybzynski 2000). One of McHarg's major foci was the connection between design theory and ecology. His book title illustrates that human intervention in the environment needs to be integrated with natural processes. 'Design *with* nature' can be seen as a normative statement, but is above all a loving perspective on life. Even if authors have different views concerning whether design with nature is in

1 In many cases this knowledge in action is considered as tacit knowledge (for example Polanyi 1966, Schön 1983, Collins 2010).

parity with the normative desire for sustainable urban development and the green city, it shows a faith in life on earth and a conviction that mankind has the ability and will to understand and adapt to their given natural prerequisites.

McHarg develops a spatial systemization for the collection and description of existing conditions and the values inherent in the landscape for future design; his is more of an empirical and investigative method than a normative approach. The values noted bridge all sectors of society. Such site- and user-related information plays a significant part in the foundation on which landscape design, including cities, is constructed. McHarg's methodology deals to a large extent with scientific knowledge as a basis for design. It is important to emphasize that the information is collected throughout the whole design process. The existing values provide a kind of framework and facilitate the ongoing design process. It would be a mistake to view this approach as mechanistic or positivist and as inherently opposed to the design process as Thompson (2009) argues. It is, rather, a systematic process for the gathering of information to be used for other modes in the design process (Murphy 2005). An indication of this is the McHarg et al. statement (1998: 357): 'We must do more than sustain the planet; we must design regenerative communities and landscapes'. Regenerating communities requires synthetic knowledge, but as importantly it depends on concern and maintenance – those values which constitute the core of what we will call the gardener's vision of the living landscape. It should also be emphasized that McHarg's approach does not concern methods of preservation and conservation (as in Florgård 2009) but is rather a way of integrating new interventions with existing values in an informed manner.

If McHarg's book *Design with Nature* is mainly seen as a methodology, Ann Whiston Spirn's book *The Granite Garden* (1984) extends method to theory. The title is a metaphor for the city, seen as a rocky landscape moulded by necessity, but managed and designed by people. Spirn systematically explores the different natural processes that city building must take into account. She also addresses how people interact with their physical city environment. With the aid of this exposition, city planners are given theoretical foundations for the implementation of the methodology developed by McHarg in *Design with Nature*.

As a product of one or many people's efforts, design means the creative and heuristic activity of producing proposals for the spatial environments of the future. The design task often consists of producing a design concept, which may be understood as a solution or several alternative solutions for the development of a site in accordance with a site-adapted goal. These design processes are interconnected with the processes of spatial planning as the latter defines several of the prerequisites for detailed design and a development project in the detailed scale of development projects can lead to adjustments in the spatial planning process. Design can also be expressed in the management and maintenance processes of urban and rural development: design through maintenance. By directed maintenance efforts in the outdoor environment, the design process occurs over a long period of time, such as in the development of rows of trees along the boulevards or esplanades of a city.

Nature as Undeveloped Land

Let us go on to explain what we mean by the concept of design with nature. In Western cultures, *nature* conventionally means *undeveloped land*, that is, where the landscape is not dominated by houses or by roads. This perception of nature involves not only wilderness but also the undeveloped terrain of cultural landscapes. In many cases, it can also mean gardens. A related and equally pragmatic view of nature can be exemplified by a statement made by the landscape architect Sven-Ingvar Andersson (1965): 'We are surrounded by phenomena that we have not created and which have a different life and a different character than the social world: trees, flowers, grass, rivers, hills, clouds. For centuries they have attracted exploration and been the object of our reverence. They have been a joy to us. In imagination, we have used them to reflect our moods. And of those, we have built up an idea, which we have called nature'. The majority of the phenomena mentioned by Andersson may be called 'nature as undeveloped land'.

According to Andersson's very common perception, nature is seen as a backdrop to human action. Interpreted this way, the constructed environment is not considered to be a part of nature. This viewpoint is problematic because it detaches human activity from nature. The consequence is that the impact of human activity on the landscape is only weakly understood. Some may think that the important perspective is cultural construction, that is, that everything is understood and constructed through the lens of human thinking and that human activity leads to representations of landscapes. But the small creatures that make up mankind should understand their role on earth. Natural processes are going on everywhere and flow through the biosphere, and natural activities and thinking are mankind's contributions to those powerful processes. We are obligated, therefore, to connect to a broader understanding of nature. We are attempting to develop an integrated approach which ignores the theoretical division into humans and nature.

The 'Laws of Nature'

Design with nature can also be understood as the smooth and harmonious management within the realms of that which can be described by the constraints or requirements nature imposes upon us: that is, that which cannot be otherwise. The concept the 'law of nature' is here aiming at the common scientific expression used in natural sciences. It represents 'regularities' in nature such as gravity that has been found in science. These laws or regularities should be seen as provisory until science comes up with even more precise descriptions, or finds new paradigms.

Within the field of planning of landscapes there is need for interdisciplinary approaches and the science dealing with spatial planning is eclectic in its character. Therefore, planning is influenced by paradigms evolved in natural sciences, social sciences as well as the humanities. The tasks in practical planning demand pragmatic approaches and necessarily lead to the invitation of pluralistic views: design activities that are informed by natural requisites, constraints and requirements. If we do not succeed in our attempts to design in harmony with these requirements, we risk testing the limits of nature to tolerate human activity. It is, for example, a hazardous enterprise to keep the sea from flooding low-lying coastlines, as in the Netherlands, but the Dutch are well aware of the risks they take. Poorly executed design will be inefficient and troublesome to all involved and can lead to higher or unsustainable costs, among other consequences.

Nature is not negotiable. There are many negotiable subjects in spatial planning and design that offer options and that can be dealt with in numerous ways. On the other hand, there are subjects dependent on natural systems or the laws of physics that do not offer such options. Rainwater flow can serve as an example. Rainwater flow follows the laws of gravity and human design has to accept and respond to it. Human designs are characterized by alternative proposals, options, negotiations, choices, priorities and decision-making. These choices should only deal with subjects that can be handled in different ways. The determined conditions defined by the laws of physics have to be taken into consideration in landscape architecture.

Construction, buildings and other artefacts are contributions naturally shaped by human intentions, but they are not independent of the constraints and requirements imposed by nature. The point to consider is whether human action turns its back on natural processes without consulting knowledge and competence to reflect upon its interaction, or whether human action is to be characterized as designing *with* the support of the same processes, to learn and benefit from them. Views on human-environmental interactions and practical design are frequently discussed in landscape architecture theory. The various alternative approaches, apart from severe anthropocentrism in design, are relevant to our gardener's view (for example, Thwaites et al. 2007, Thompson 2009).

Three Categories of Nature as Landscapes

A closely related understanding of nature as undeveloped land is the interpretation of nature as landscapes. Landscape historian John Dixon Hunt (1992) elaborates on an expression by the Roman writer Cicero and addresses the three natures. In this definition, nature is more easily understood as different types of landscape. Today we would probably designate these 'natures' as wilderness, cultural landscape and the garden and we understand that they have been shaped by various degrees of human intervention. The garden is the pinnacle of a human-influenced landscape considered a part of nature, a place '… where the utmost resources of human intelligence and technological skills were invoked to fabricate an environment where nature and art collaborated' (Hunt 1992: 4, Hedfors et al. 2008). The theoretical analysis of the three categories of nature are useful in the evaluation of existing patterns of land use in the purpose of reconsidering the character and substance of the relation between the constructed environment and the landscapes and natural resources of the wider region. The evaluation of existing patterns of land use is one of the purposes of this anthology.

The First Category of Nature

The three categories of nature help us understand the gardener's perspective. The relevance of the first category of nature for the gardener's vision may be, for instance, the genetic traits of the plants in the garden. Spontaneously established plants can provide an example, if they decide to favour them through maintenance. The gardeners incorporate these aspects in their idea of configuration and wilderness thus becomes one of the colours on the designer's palette. Another aspect of wilderness in its original sense is that it hardly exists. A common argument for this is that mankind nowadays influences all parts of our planet by air and water-borne pollution. An even more striking argument is the regulations, agreements and comprehensive plans established by authorities cover almost all land, on different levels. These plans, to a certain extent, regulate the use of land and water and thereby add human intent to all spatial areas. National parks have intentionally been preserved and thus wilderness has been incorporated in the realm of cultivation.

The Second Category of Nature

The second category of nature lacks a cohesive idea of configuration when compared to the third. Different actors, their individual interests and specific use of the site, and the lack of coordination and a comprehensive idea characterize the cultural landscape. A certain amount of coordination is imposed in certain cases, for instance, in borders between properties or to allow roads and other shared interests to function properly. But this coordination is not the result of a cohesive, conscious design; it is rather the consequence of necessary practical actions. Without a coordinated policy it is more difficult to predict the consequences that different activities have on the landscape. This fragmented landscape of borderlands requires negotiation between neighbours concerning how borders between properties are managed. One neighbour may wish to keep a tree to ensure privacy while the other may wish to remove it to increase direct sunlight.

The Third Category of Nature

For the definition of the third category of nature, we can turn to Hunt (1992) who describes the garden as the abode of the landscape gardener, skillfully weaving together the wishes of people and other interested parties and responding to them in a creative way, working within a coordinated idea of configuration. Unlike the landscape architect's fixation on a single set of blueprints, the gardener implements a new configuration proceeding from a consciously structured idea that develops over time by virtue of their constant presence on the site.

The City: Second or Third Nature?

The historian John Stilgoe (1982) attempts to define the cityscape, as opposed to the landscape, as places where mankind wholly dominates the land, when they shroud it almost completely with structure and moulded space. Stilgoe regards the landscape as mainly rural – an interpretation that corresponds to the Latin language's use of *paesaggio* (Italian), *paysage* (French) and so on. However, the European Landscape Convention (CoE 2000) is a promising convention, which takes into account the holistic landscape concept that bridges the three natures and that is also valid for urban landscapes. In the European Landscape Convention cities, as well as national parks undisturbed by man, are seen as parts of landscapes. This conceptualization is in common use within the discipline of landscape architecture (ECLAS 2011).

A reasonable approach for landscape design is an integrative perspective on human intention in nature. The integration means that human intention and cultural or social construction are parts of nature. Human intention in the wider landscape becomes stronger and stronger from within the wilderness, to within the cultural landscape, and eventually to within the garden. Hunt's (1992) garden as the third nature contains the greater number of intentions: it has the highest 'density of intentions', while Stilgoe's cityscape is either an example of the second or the third nature, depending on the inhabitants' ability to organize a sustainable urban development in a systematic manner. The development of urban landscapes often shows a relationship to the second nature, the cultural landscape; an *ad hoc*-based development of cities without an articulated strategy.

The three natures stand for an escalating degree of intention and the city can be defined as the landscape, which harbours the greatest degree of human intention, as Hunt's garden and Spirn's *Granite Garden* suggest. Nonetheless the contemporary city's development most often lacks an overriding strategy and therefore may be referred to as an ungoverned second nature, characterized by multiple expressions of individual interest. By contrast, the third nature – the garden metaphor – requires the utmost application of human intelligence and technological skills to fabricate an environment where nature and art collaborate; but to achieve this end requires an overriding strategy. The driving forces of change – whether financial, environmental or social in origin – are blind: little thought is given to the effect on the development of the landscape. In the end, we face a choice: the city landscape can either mainly be an outcome of social processes lacking an all-encompassing spatial strategy – an example of the second nature, or mainly be constituted of planned spaces – an example of the third nature. The gardener's vision connects to a broad understanding of nature by interpreting the three categories of nature from an integrated standpoint.

Bridging the Nature/Culture Dichotomy

A strictly limited definition of landscape is put forward by Stilgoe (1982), when he distinguishes chaos from intent, wilderness from human-imposed order. According to this position, man must choose between landscapes; integration is impossible. By contrast, our understanding of design with nature is based on the different meanings of landscape. An inclusive perspective of culture as part of nature is equally important. Stilgoe (1982) states that landscapes in the meaning of cultural landscapes always display a fragile equilibrium between natural and human forces. The dichotomization that occurs by seeing human forces *apart* from the natural, as Stilgoe does, is predominant in Western culture. This separation can be observed for instance in the organization of national institutions. For example, the board of nature preservation is separated from that of cultural heritage preservation. The separation is also found in the academic discipline of geography that was divided several decades ago into natural or physical geography and human or cultural geography. Another academic discipline, landscape architecture, has so far avoided this division and develops a more inclusive approach, probably because of its practical orientation. 'Once we can accept that the city is as natural as the farm and as susceptible of conservation and improvement, we work free of those false dichotomies of city and country, artificial and natural, man versus other living things' (Lynch 1981).

To bridge such separated perspectives and merge them in practice requires a landscape architecture that pragmatically develops the site based on its existing values and opportunities. Our gardener's perspective, like human cultures in general, can develop different approaches: design with nature or dissociation from nature. The preposition 'with' as the third component of the concept has a significant role in relating design and nature to each other. With alternative prepositions that lie between design and nature, the meaning of the concept will change dramatically. If we suggest 'design *for* nature', our activities would become representative of nature – a non-anthropocentric view. If we instead propose 'design *as* nature', our activities would expressly be based on reliable knowledge; otherwise they would be dedicated to superficial mimicry – a kind of stylized romanticism. If we state 'design *after* nature', the field of knowledge would be dedicated to adjustments of human errors that have come to human awareness by observation, such as pollution, or by the deficient management of natural disasters. The movement of ecological modernization (for example Mol et al. 2009) may be considered to follow that path (Table 9.1, Hedfors 2009).

Another preposition to examine is 'of' as in 'design of nature'. Pure design founded on conscious choices refers to the third nature: the garden. The intention to create a 'design *of* nature' is to be defined as a version of this third nature (Table 9.1).

It implies a dominant attitude, that a single species, the human race, should be able to develop settings on its own. The philosopher Arne Naess (1989) states: to distance oneself from nature and the 'natural' is to distance oneself from a part of which the self is built up. The preposition 'with' associates, as the others dissociate.

Table 9.1 Concepts addressing different positions to our environment in design based on attitudes detached from or engaged with nature

Concept	Meaning	Engaged with nature	Detached from nature
Design with Nature	Integration	X	
Design for Nature	Representation		X
Design as Nature	Mimicry		X
Design after Nature	Reparation		X
Design of Nature	Domination		X

An Example: Oslo Coping with Nature

Urban settings are seen as agglomerations in the wider landscape and design with nature in cities is a challenge, one which the local authorities of Oslo have met in several ways. The large recreation areas, or '*Markas*', are mainly unexploited and are striking examples of how a city can cope with nature seen as undeveloped land. But also the small seemingly trivial remnants of nature, used as everyday relaxation areas and natural playgrounds, are of great importance. Preservation of such vegetation provides aesthetical, social, functional, biological, historic and economic advantages. The promoting force of urban sprawl that threatens most cities correlates with such preservation, as new developments search for 'untouched' positions in the landscape. Oslo local authority has so far dealt with this force. However, urban sprawl lacks a design with nature strategy and a holistic perspective of the landscape. As urban sprawl is characterized by a more or less uncontrolled exploitation, there are no features in the existing landscape that are valued or preserved and that might contribute to the future place identity. Thus, design with nature is a perspective of high significance in challenging densification strategies as well as urban sprawl, and in involving the patterns of apparent nature and greenery in city regions. Design with nature involves keeping your ear to the ground both when compacting and expanding the city, or when developing satellite agglomerations.

In landscape architecture there are always actual situations and existing conditions to take into account, and the situation of a *tabula rasa* is seldom encountered. In a first planning stage, decisions must be taken as to how an area should be used: whether it should be preserved as it is or whether it should be modified and, if so, in which direction. To consider values of preservation defined by stakeholders is the opening step in landscape design (Florgård 2009). Our new ideas and proposals ought to be related to existing values of various kinds at a site. The challenge is to incorporate existing values, that is, to preserve them and to develop the site-specific potentials in accordance with the natural processes on site as new concepts of design are added.

The Metaphor: The Gardener's Pragmatic View and Design through Maintenance

Mankind has played the role of gardener from the very beginning. The gardener muddles through, but to succeed in acting that way they are obliged to continuously use their experience and develop their knowledge and competence regarding natural processes, human-environmental interrelations, aesthetics and so on. The role of the gardener was sufficient until human beings developed tools that could cause massive damage to nature, mankind included; then ecological knowledge and competence did not exist or was neglected.

Design with nature fosters a view of nature proposed, among others, by Spirn (1995: 254), consisting of creative- and life-sustaining processes, which connect everything in the physical and biological worlds, including humans. Although ideas of nature are products of culture and vary from people to people, place to place and period to period, this inclusive view of nature corresponds with the gardener's vision. The gardener's vision simply does accept any theoretical separation between human beings and nature, or between artefacts and environment. Instead gardeners deal pragmatically with their landscape as a whole. They are not in need of any instrumental step-by-step procedure to understand their environment, as an anthropocentrist requires. The emergence of human ecological consciousness is a philosophically important idea, according to Naess (1989): it signifies the arrival of humankind as a life form which is capable of understanding and appreciating its relations with all other life forms and to the earth as a whole. The gardener's strategy based on design through maintenance emphasizes a continuous approach along with the competent application of knowledge. The perspective is used in the development of LandACT (Landscape Architecture City Theory in Hedfors et al. 2008).

In the garden – in a broader sense meaning the entire landscape including human settlements – there is recognition of constructedness and an attitude of beneficial management, as well as an acknowledgement that certain non-human phenomena are beyond human control (Spirn 1995). That dynamic vitality in the self-regulating processes of life is what the gardener metaphor seeks to capture. The gardener's perceptiveness is suited to interventions in all three categories of nature: in all landscapes, including cities. The ways in which landscapes are shaped and the involvement of stakeholders are common in spatial planning and design, but the gardener's sensitivity is missing; their user orientation, their sensitivity to the growth of vegetation, to water flows, and to other natural processes is not present.

Concluding Remarks

Initially, we posed the question: what does a practical approach within city building imply where densification and uncontrolled spread of construction is avoided, where the prerequisites of the existing landscape and its innate life force is taken as a starting point? We have used the gardener as metaphor to re-examine

Figure 9.2 The sculpture of Nils Ferlin is 'constantly visiting' the open air and the living landscape of a park in Filipstad, reminding us that the gardener's view deserves to be brought into the daylight and applied to the whole landscape

Source: Photo by P. Hedfors, 2011.

the concept of design with nature. The existing conditions of a site are proposed to be the starting point of an intervention. This proposal can be repeated many times. Additionally, a non-anthropocentric approach is central as is an understanding of the living landscape and its dynamic processes: vegetation growth, water flows and so on, as well as their interactions with people. The proper competence, symbolized in our chapter by gardening is an important competence that should accompany the early stages of city planning and other development projects. Finally, design through maintenance aims at the landscape architect's ability to successively design a place with the help of the dynamic processes at the site and manage them intentionally. The landscape architect's presence in city development, or the presence of his/her view, contributes to the long-term management and to sustainable urban development.

Our story began with poor Nestor situated underground surrounded by concrete, neon, cars, and polluted air. Another well-known bronze sculpture in Filipstad, Sweden, depicts a famous Swedish poet from the same town, Nils Ferlin (Figure 9.2). This sculpture is situated in the open air surrounded by flowers, trees, flowing water, fountains and sunlight. This location would seem more familiar to Nestor. The marketing manager of the parking garage has succeeded in creating tension, contrast and debate by inviting Nestor underground. Art is a significant contribution of mankind to nature that evokes human thought. Creativity is one of our leading forces, but we propose that the gardener – and especially the design philosophy that we metaphorically call the gardener's view – should be moved above ground.

Bibiography

Ahrland, Å. 2005. *Den osynliga handen: Trädgårdsmästaren i 1700-talets Sverige* (*The Invisible Hand: The Gardener in 18th Century Sweden*). Alnarp: Sveriges lantbruksuniversitet.

Andersson, S.I. 1965. Review: Modern gardens and the landscape, by Elisabeth B. Kassler. *Havekunst*, 8, 158.

Berg, P.G. (ed.) 1993. *Biologi och bosättning: Naturanpassning i samhällsbyggandet* (*Biology and Settlement, in Swedish*). Stockholm: Natur och kultur and Institutet för framtidsstudier.

CoE 2000. *European Landscape Convention.* Strasbourg: Council of Europe.

Collins, H.M. 2010. *Tacit and Explicit Knowledge*. Chicago, IL: The University of Chicago Press.

ECLAS 2011. *Landscape Architecture: The European Dimension* [Online: European Council of Landscape Architecture]. Available at: www.eclas.org/landscape-architecture-european-dimension.php [accessed: 18 August 2011].

Florgård, C. 2009. 'Preservation of original natural vegetation in urban areas: An overview', in *Ecology of Cities and Towns: A Comparative Approach*, edited by M.J. McDonell et al. Cambridge: Cambridge University Press, 380–98.

Hall, L. 1995. *Olmsted's America: An 'Unpractical Man' and His Vision of Civilization.* New York: Bulfinch Press.

McHarg, I.L. 1969. *Design with Nature.* Garden City, NY: Natural History Press for the American Museum of Natural History.

McHarg, I.L. and Steiner F.R. (eds) 1998. *To Heal the Earth: Selected Writings of Ian L. McHarg.* Washington, DC: Island Press

Hedfors, P. (ed.) 2009. *Urban naturmark i landskapet: En syntes genom landskapsarkitektur*, festskrift till Clas Florgård (*Urban Nature in the Landscape, in mixed languages*). Rapporter, 3, Swedish University of Agricultural Sciences, Dept. of Urban and Rural Development.

Hedfors, P. and Granvik, M. 2008. *Landscape Architecture Theory: A Source for Interdisciplinary Sustainable Design.* Proceedings of the *Environscape: A Manifesto*, 2nd blu+verde International Congress, edited by L.M.F. Fabris (Milan: Politecnico di Milano, Faculty of Architecture, Department BEST, 23–24 October), 327–36.

Hunt, J.D. 1992. *Gardens and the Picturesque: Studies in the History of Landscape Architecture.* Cambridge, MA: MIT Press.

Lang, J. 1987. *Creating Architectural Theory: The Role of the Behavioral Sciences in Environmental Design.* New York: Van Nostrand Reinhold.

Lynch, K. 1981. *A Theory of Good City Form.* Cambridge, MA: MIT Press.

Mol, A.P.J., Sonnenfeld, D.A. and Spaargaren, G. (eds) 2009. *The Ecological Modernisation Reader: Environmental Reform in Theory and Practice.* London: Routledge.

Murphy, M.D. 2005. *Landscape Architecture Theory: An Evolving Body of Thought.* Long Grove, IL: Waveland Press.

Naess, A. 1989. *Ecology, Community and Lifestyle: Outline of an Ecosophy.* Cambridge: Cambridge University Press.

Polanyi, M. 1966. *The Tacit Dimension.* Garden City, NY: Doubleday and Co.

Roper, L.W. 1973. *F.L.O.: A Biography of Frederick Law Olmsted.* Baltimore, MD: Johns Hopkins University Press.

Rybzynski, W. 2000. *A Clearing in the Distance: Frederick Law Olmsted and America in the 19th Century.* New York: Scribner.

Schön, D.A. 1983. *The Reflective Practitioner: How Professionals Think in Action.* New York: Basic Books.

Spirn, A.W. 1984. *The Granite Garden: Urban Nature and Human Design.* New York: Basic Books.

Spirn, A.W. 1995. 'The authority of nature: Conflict and confusion in landscape architecture', in *Nature and Ideology: Natural Garden Design in the Twentieth Century*, edited by J. Wolschke-Bulmahn. Dumbarton Oaks Colloquium on the History of Landscape Architecture (18), Washington, DC: Dumbarton Oaks Research Library and Collection, 249–61.

Stilgoe, J.R. 1982. *Common Landscape of America: 1580 to 1845.* New Haven, CT: Yale University Press.

Thompson, I.H. 2009. *Rethinking Landscape: A Critical Reader*. London: Routledge.

Thwaites, K. and Simkins, I. 2007. *Experiential Landscape: An Approach to People, Place and Space*. London: Routledge.

WCED (World Commission on Environment and Development) 1987. *Our Common Future*. Nairobi: United Nations Environment Programme.

PART III
Plans, politics and outcomes

Chapter 10
How Green is Oslo? Urban Development, Green Policies and the Making of a Green Capital City

Guttorm Grundt

Oslo as a Green European Capital

Major environmental award competitions and indexes rank Oslo high among European 'green' capitals. Oslo won the European Sustainable City Award in 2003, together with Heidelberg and Ferrara, and qualified for the finals in the European Green Capital Award competition in 2008, finishing sixth out of 35 participating European cities with more than 200,000 inhabitants. Oslo was also ranked third out of 30 European capitals in the European Green City Index 2009, behind Copenhagen and Stockholm. The competitions were organized by the European Union DG Environment, whilst the ranking was made by the Economist Intelligence Unit, sponsored by Siemens. The EU commitment to organize an annual Green Capital Award is a following up of its *Thematic Strategy on the Urban Environment.*

The assessments were broadly based upon a set of common indicators of environment and sustainability, initiated by the European Sustainable Cities and Towns Campaign, and further developed with support from DG Environment. The two competitions still differed somewhat in focus. The European Sustainability Award focused mainly on sustainability issues central to the Local Agenda 21 process. The European Green Capital Award focused more strictly on environmental issues, applying a set of ten environmental indicators. (DG Environment/Ambiente Italia Report 2010). The European Green City Index used many of the same environmental indicators as the Green Capital competition, but focused also on energy and building quality (Economist Intelligence Unit 2009).

As can be observed from the context and description of the three competitions, 'green' is not to be taken literary to mean the greenest looking, or the most vegetation-dominated city. 'Green' is a measure of the environmental quality of a city in terms of a set of specified indicators. Environmental quality, therefore, is a vital element of urban sustainability.

Figure 10.1 Oslo received the European Sustainable City Award in 2003; logo pictured

Environmental Assessment of Oslo Today

Oslo's score on the indicators compared to the eight 'greenest' European cities ranged from top position on green structure, sustainable energy and climate gas emissions, to middle position on air quality, noise, land use, transport and environmental management, to poor on waste, water and waste water. Some of the rankings came as a great surprise to Oslo's vigorous environment lobby, reflecting the differences in environmental challenges and priorities between Norway and continental Europe. A brief summary of Oslo's relative environmental state and performance on the different indicators reveal a picture of Oslo as a relatively well managed major 'green' capital city, rich in natural resources, forest, water and renewable energy, but also with a standard of living stimulating relatively high rates of consumption of both water, energy and materials.

Environmental management: Oslo has a working environmental management system. An environmental policy called Urban Ecology Programme adopted by the city council gives direction to the different strategies and long-term plans of the relevant agencies, urban districts and other parts or the municipal organization. Environmental performance is reported annually, and environmental status and policy revised once per city council period. All municipal agencies and units are obliged to develop a programme for environmental efficiency of their own administration, become eco-certified and practise ecological procurement practices.

Green space: two-thirds of the area within Oslo's municipal boundaries is designated forest, parks, lakes and rivers. Oslo has a high ratio of population to public green spaces, with 44 square metres for each inhabitant. Oslo has also a high accessibility of public open areas with 94 per cent of the population living within 300 metres of a public green area. In 2009 the existing *Marka* or forest encompassing Oslo was officially designated an ecological preserve and outdoor recreation park. The extensive forested area is protected from urban development.

Climate/CO_2 emissions: Oslo has very low climate gas emissions with 2.1 tons per inhabitant (CO_2 equivalents). The low emissions are mainly due to a high percentage of renewable energy consumed by the city (64.8 per cent from hydroelectric power). Emissions from fossil fuels in transport represents 64 per cent of the emissions, industry 17 per cent, private households 9 per cent, waste (methane) 7 per cent and the last 3 per cent from other sources. On the other hand, Oslo has not managed to reduce its climate gas emissions from the 1990 level.

Buildings/energy efficiency: Oslo has a high-quality building stock and no slums. Oslo has high energy efficiency buildings standards, and energy consumption in residential buildings is relatively low. Parts of the energy efficiency investments have been financed with loans and grants from the city's own climate and energy fund, which also supports the replacing of oil burners with heaters based on renewable energy sources.

Air quality and noise: during the winter, Oslo has seasonal problems with air quality, especially along the main roads. Annual mean levels of Nitrogen dioxide (NO_2) exceed the EU limits in the central parts of the city, caused by emissions from motor vehicles, both petrol and especially diesel fuel, and aggravated by winter temperature inversions. Oslo no longer exceeds the EU directive maximum levels of Particulate Matter/dust (PM_{10}), due to the introduction of a fee on the use of studded tires, improved road maintenance and municipal grants to replace polluting old wood burning stoves. Noise abatement measures have reduced the number of inhabitants exposed to noise, and Oslo is working especially to designate a number of areas in the city as 'quiet areas'.

Land use: Oslo has an active policy against urban sprawl within the municipal boundary, and the built-up area only takes up one-third of the total area. The land use policy builds upon the principle of densification on 'grey' areas within the built-up area, and especially around the nodes on the public transport system. The potential for further densification is high, as the overall population density is relatively low with ca. 1,300 inhabitants per square kilometre and 3,800 in the built-up area, which is about only half of the density of Stockholm. Oslo's strict densification policy is only partly practiced by the surrounding municipalities in the Oslo region.

Transport: 89 per cent of the population live within 300 metres of a public transport stop in Oslo and rail-based public transport (metro, trams and trains) carries almost two-thirds of the passengers within the city limits. Public transport has a 45 per cent share of all motorized travel in Oslo, and 31 per cent in the region. Whilst the population in Oslo has increased by 17 per cent during the last ten years, public transport has increased by 28 per cent, and car traffic has increased by only 10 per cent. Oslo has relatively high share of trips on foot for trips under 5 km (50 per cent), and a lower than average share by bike (7 per cent) compared to the average finalist. The low share of biking can partly be related to unfavourable topographical conditions and snowy winters, and partly to the limited amount of dedicated cycle tracks (0.42 metres per inhabitant).

Waste and consumption: Oslo has a relatively large ecological footprint requiring eight hectares of productive area per person to sustain the lifestyle and standard of living of its inhabitants. This is four times the global average, but still only half of that of the USA. Oslo produces ca. 380 kilos of household waste per inhabitant, of which 65 per cent is incinerated for energy recovery. The energy is used to supply the city's district heating system and 30 per cent of waste material is recycled. Glass and metals are collected at over 600 return points. Furniture, household appliances and redecoration waste and so on can be delivered at one of the city's return stations for sorting and reuse. Garden waste can be delivered at one of the city's composting stations. An improved cycle-based waste treatment system is to be completed in 2012, adding the home sorting of plastic and wet-organic (kitchen) waste to the sorting of paper and residual waste for collection.

Water consumption and waste water: Oslo takes its drinking water supply from the forest lakes surrounding the city. The water is treated in two modern water treatment plants and piped to the city: 20 per cent of the drinking water is lost in pipe leakages. Only 30 per cent of the households use water metering. This contributes to Oslo's high water consumption with 446 litres per inhabitant per day. The sewage water is piped to two modern waste water treatment plants before it is discharged into the fjord at deep waters. The removal of nitrogen from the treated waste water is suboptimal (70 per cent). During heavy rain the storing capacity of the sewage pipes and ducts leading to the treatment plants is exceeded, and the overflow is discharged into the fjord.

Historical Development of Oslo: An Environmental Analysis

Oslo's development has a close resemblance to that of the other eight major European green cities, but also some very specific national and local characteristics. In truth, Oslo's score and rating reveal that both differences in natural settings and cultural traditions, along with corresponding differences in political priorities, explain much about Oslo's specific 'green' profile. We shall now take a closer look at the different phases of Oslo's 'green' development, and the way in which the various

Figure 10.2 'Oslo in the middle ages'

Source: Photo by R. Aakvi, by permission of Oslo Museum.

economic, social, legal, cultural and political forces and factors have influenced this development through time. These will be illustrated in brief through following the winding trail of Oslo's environmental history. The development of Oslo has gone through several phases in response to varying environmental qualities and challenges. These are described briefly in the following phases, supplemented by a description of measures and actions taken to preserve the qualities, compensate for qualities lost and meet or reduce the challenges.

Phase 1: The early 'green' capital town, 1000–1850

Oslo was founded around the year 1000 and received its royal charter in 1050. The city was made Norway's capital city by King Håkon V Håkonson in 1299. Nonetheless, the capital function was in practice moved to Copenhagen in 1375. As a consequence, Oslo remained a small provincial town for nearly 400 years; in fact during its first 700 years, Oslo's population averaged around 3,000 inhabitants, and as late as in 1769 the first registered census counted only 7,469 inhabitants.

Although Oslo was restored as the nation's capital in 1814, the Swedish King and his government continued to rule Norway from Stockholm for another 90 years. Consequently, Oslo's growth was not so much a result of its status as a capital city, but accrued from the growth of trade and the beginnings of industrialization. Starting around 1840, the pace of the city's development increased dramatically.

In environmental terms, old Oslo was a small, relatively self-contained town. Most of the resources needed to sustain urban life were produced in or close to the town. Local wells, brooks or rivers provided drinking water, and food was provided by farmers and fishermen in the region, or by the town dwellers themselves. Many of the households kept their own hens, pigs and vegetable gardens. Local clay was used to produce bricks and tiles, and building timber and firewood were supplied from the surrounding forests. Faeces and manure were used as fertilizers for local food production, and other waste was largely material or energy recycled. Land transport was on foot or by horse, and sea transport by boats powered by oars or sail. The use and reuse of short travelled natural resources kept the inhabitant's ecological footprints small and sustainable.

Old Oslo may seem to us like an environmental paradise, but it was not quite. Nearby woods were stripped for firewood, and open fires and sewers caused local air pollution. Latrines and waste water polluted the local groundwater, and repeated town fires destroyed valuable resources. Run-off from the local aluminium mining works and tanner's shops polluted the nearby Alna River. But this was all on a small scale and of no regional or global consequence. After a series of large city fires the king decided to move the city closer to the city castle, and the modern baroque plan organized a city with wide airy streets, stone or brick buildings and cobbled paving.

Phase 2: The Industrial City, 1850–1930

Oslo's population growth exploded from the middle of the 19th century, with a sevenfold increase in population to 330,000 in 1925. Industrialization was the main driving force, initially linked to the supply of sustainable hydropower from the area's largest rivers, the Alna, Akerselva and Lysakerelva. Another major driving force was Oslo's capital status generating the development of a new national government and its institutions. Government and industry attracted trade and finance, and Oslo soon developed into Norway's most important port for both imports and exports as well. The economy thrived and the new wealth changed the character of Oslo from a small provincial town, but its path to becoming a cosmopolitan European capital could begin only after Norway gained full sovereignty from Sweden in 1905. The strong population growth required housing and necessary infrastructure for energy, water, sewage, waste, transport and recreation. This required more governmental management, and Oslo

Figure 10.3 Oslo's industry was founded on the water power of the Aker River

Source: Wood engraving by by C.E. Baagøe, 1867 by permission of Oslo Museum.

developed municipal administration and services to cater for the needs of the growing population. The city growth soon spilled over its administrative borders, and subsequent city extensions required more land for building. But the pressure from the massive growth often put a strain on the modest public administrative capacity to contain an urban development dominated by uncoordinated initiatives from builders and developers exploiting the market.

Oslo's growth and progress had its environmental costs, mainly caused by increasing air and water pollution. Nonetheless, bold decisions and strategic investments of this era also contributed to Oslo's green qualities to be enjoyed by generations to come. Some of the 'solutions' resulted, however, in even bigger remediation costs for subsequent generations. The increase in air pollution was mainly caused by the increasing use of fossil fuel for energy in the growing city. The use of firewood and energy from the city river falls was partly replaced by imported coal in the latter part of the 19th century. Coal was used as an energy source for housing and fuelled steam engines for industrial production, as well as rail and ship transport. Coal was also processed to produce gas. The first gasworks was established in 1848 primarily to produce gas for street lighting, but the use of gas soon became popular for other uses. The black and smelly fumes from the noisy gasworks represented a serious environmental problem for parts of the city. The city established its own electric energy company in 1891, and Oslo's first hydroelectric power station was built in Maridalen in 1900.

Water pollution was the other main environmental challenge in this period. The pollution of wells, brooks and rivers used for drinking water, together with scarcity of water for a growing population, resulted in a need for clean and sustainable drinking water resources in adequate supply. The solution was to take the drinking water from further upstream the Aker River, and later Maridalen Lake, and pipe it to posts and buildings in the city. To secure the water resources and its precipitation area, the city bought up large parts of city forests. Waste water from kitchen sinks and other sources was originally led to gutters and brooks together with rainwater. The priority of hygiene before environmental protection resulted in the introduction of the water closet, and, as late as in the 1930s, the sewage from 75,000 inhabitants was discharged from sewage pipes and ducts into the streams, rivers and the fjord, with only mechanical treatment. Sludge from the 'treatment plants' was towed out on large barges and dumped into the fjord. The strong pollution of the streams and rivers led to the encasing of many streams to reduce the odours and increase safety for children.

Solid waste was for many years still used as a valuable organic resource for agriculture. Household waste, including faeces, was composted and used by vegetable gardeners and farmers outside the city. Organic kitchen waste was sorted, collected and sold to pig farmers. A growing amount of waste from a growing population created a need for control and other additional solutions, however, and consequently a municipal waste company was established in 1897. Larger landfill sites were unpopular with the public and hard to find. In 1908 the bulk of the city waste was shipped by barges to the *Langøya* islands in the inner fjord and deposited in the sound between them, a solution which lasted till 1949. The filling was another blow to the water quality in the inner fjord, and a feeding station for flies and seagulls.

The expansion and transformation of Oslo from a town to a city increased the distance to nature, and created a desire, and need, for green nature in the city itself. The city responded by establishing a special tree planting service in the engineering department in 1875, upgraded to the City Park Agency in 1916. The parks structure of Oslo was, however, originally more a result of chance than design. The first city parks, St. Hanshaugen and Stensparken, were small; they were created on steep, nearly unbuildable land already used by the public for recreational purposes. The famous Frogner Park in the west was originally acquired for use as a cemetery and building land. Social design based on access to landscape found form in the large Ekeberg hill in the east, where 210 ha. of land was acquired to secure open air recreation for the working class from the eastern parts of the city. The plain on the top of the hill is today the venue of Norway Cup, the world's largest soccer cup for children. A plan for the rehabilitation of the lower part of the Aker River to a river park was adopted in 1917, but it would take 70 years till the whole length of the river was rehabilitated to river park quality.

Figure 10.4 Waste logistics at Long Island's sea-fill

Source: Photograph A.B. Wilse, ca 1930; by permission of Oslo Museum.

Phase 3: Urban Expansion and the Motor Car Age, 1930–1970

The key to address the problems and challenges of the fast growing city was the introduction of comprehensive urban development planning, piloted by Oslo's first city planning director, Harald Hals. Hals published Oslo's first master plan 'From Christiania to Greater Oslo' in 1929, providing a strategy for directing and mastering both the 'modernization' and transformation of the inner city to a well-functioning city centre, and the city's expansion and development for the coming years. The comprehensive plan reflected the strong public concern for improving the sanitary, social and environmental conditions for the city's poor and growing working class.

The continuing influx of people, goods and capital gradually outgrew the city borders, and Oslo and the adjacent Aker municipality were joined together in 1948. This facilitated large-scale urban development to tackle the chronic housing shortage, aggravated by the war in 1940–1945. The new city master land use plan of 1950 gave directions for the development pattern and infrastructure for the city, with building land for a population of 700,000 inhabitants. Two thirds of the area within the city borders was reserved for nature and open-air recreation, mainly comprising the city forest to the north and east. New housing should mainly be built as self-contained 'satellite' townships well served by rapid rail. A network of green corridors was reserved in between the new 'satellite' townships and other built-up areas of the city, giving green access to both the city forest and the fjord. The city bought up whole farms for development, and leased or sold the land to newly established housing cooperative associations or companies. Housing development was coordinated with the extension/building of the tramway and later, metro lines, built by the public transport company Oslo Sporveier AS.

By 1933 all new and old buildings in Oslo enjoyed electric lighting. But an increasing demand from a growing industry and population soon forced the city to look outside the city borders for the supply of more electricity. The municipal energy company Oslo Lysverker AS acquired water-fall rights and dammed rivers

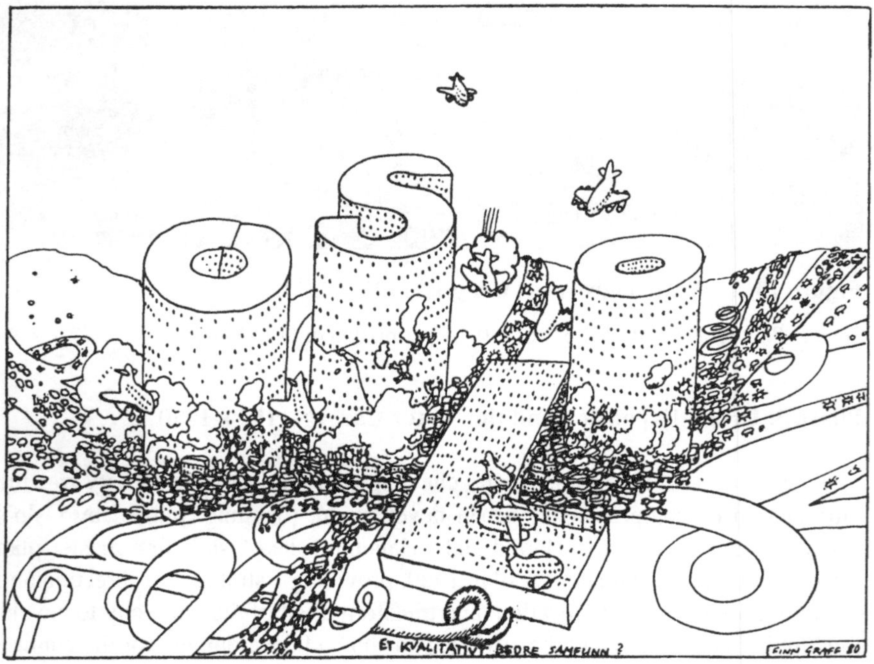

Figure 10.5 'The motor car driven dystopia'

Source: By permission of the artist, Finn Graff.

in far away valleys, built power stations and transported the energy to the city in large power lines, not without protests from the nature preservation lobby. Hydroelectric power has since been the main basis of Oslo's energy supply, still delivering two-thirds of Oslo's stationary energy, providing renewable energy at reasonable prices.

The reduction in air pollution from the cessation of the burning of coal and oil for energy production and heating was soon to be reversed by emissions from the growing car traffic. The strong increase in public transport before the war was turned to a market fall as the number of motor vehicles increased from 30,000 in 1948 to 73,400 in 1959. Motor traffic by vehicles registered in the region comes in addition to this. The city streets and roads were not built for this amount of traffic, and the capacity was hopelessly inadequate. The growth in parking capacity brought even more cars into the city. By the end of the 1960s the detrimental environmental effects of the motor age could no longer be neglected, but neither could every family's dream of a private car.

If air pollution remained a problem, water pollution from sparsely treated waste water (and solid waste) resulted in an ecological catastrophe, with poisoned rivers that soon atrophied, killing off the trout. Warnings from the health authorities were issued against swimming in the rivers and the inner fjord basin. The inhabitants were also warned of the health risk of eating fish caught in the inner fjord. Compulsory waste collection was introduced in 1932, and in 1937 a landfill was opened on *Fornebu* airport, using the filling mass to prepare an airport for the city. Other landfills took over when the airport opened in 1947. The landfill between the *Langøya* islands was finally covered with clay, sand and grass, and is today a popular island for swimming, camping and recreation.

The 1930s saw an explosion in the public interest for outdoor recreation. Parks were important enough, but for many inhabitants the most important part of the green open space was the city forests and islands. The skiing and ramblers associations marked cross-country skiing tracks and foot paths in the *Marka* forest, and built timber lodges to accommodate the sporty citizens for refreshments and staying overnight. The Oslo Fish Administration (OFA) was established to monitor and cultivate the fishing stock in rivers and lakes to stimulate sport fishing. The Oslo Regional Outdoor Recreation Council (OFF) was established by the effected municipalities mainly to protect the *Marka* forest from urban development, and the Oslo Fjord Outdoor Recreation Council (OF) was established by the boating clubs and bordering municipalities to acquire land and cater for public access to the still undeveloped beaches and islands along the 100-km-long Oslo fjord.

The Rise and Fall of the Oslo Regional Planning Office

The regional challenges of Oslo's urban development and sprawl were, up until the last war, largely focused on the need for coordinated planning and development between the City of Oslo and the surrounding Aker Local Authority. This was partly solved with the fusion of the two local authorities in 1948. Oslo's strong economic development as a centre for industry, business and public administration was, however, also a strong magnet on both capital and employment already between the wars, with urban sprawl taking place in many municipalities outside Oslo and Aker. The population increase in the municipalities surrounding Oslo was 41 per cent between 1948 and 1960, compared to 14 per cent for the enlarged Oslo city. Oslo's share of the region's employment increased from 82 per cent in the 1930s, to 88 per cent in 1960, whilst the share of the workforce in the outer municipalities working in Oslo rose from 20 per cent to 45 per cent. The regional challenges created by the ensuing urban sprawl called for efficient long-term regional coordination and planning, and a regional committee and office were established in 1950 by the Ministry of Local Government, the City of Oslo and 15 local authorities in the region. The committee studied the different challenges caused by the strong regional pressure facing Oslo and the municipalities, and wrote reports and white papers for political consideration. But the much needed comprehensive regional plan was still missing.

The solution seemed to arrive with the revised Planning and Building Act of 1965, which gave a legal basis for the establishment of formal regional planning. A regional planning council was set up in 1968, and a new regional planning office proposed its coordinated land use and transport development plan for the Oslo region in 1974. The regional development plan was never ratified by local governments, and the regional planning office was closed in 1976, not for lack of regional purpose, but by lack of political will to make it work. The main political problem was the reluctance of the surrounding municipalities to embark on close cooperation with Oslo, which in the long run might reduce their status to just another one of Oslo's urban districts. At the same time, Oslo has always been cautious about losing its independent role, with powers and economic resources as a county of its own. There have been several later attempts to revive regional planning in the Oslo region, including county planning, but with meagre results.

Although urban sprawl and the transport congestion have continued in the region, there has been successful regional cooperation on a few, but vital, areas for the region's functional and environmental performance. The regional open-air recreation councils have already been mentioned. A public transport company for Greater Oslo was established by the two counties in 1975 to coordinate and stimulate the use of train, metro, tram bus and boat traffic in the region, including common tariffs. The Oslo fjord project was established in 1977 for the research, monitoring and improvement of the water quality of the polluted fjord, stimulating the construction of inter-municipal wastewater treatment plants on both sides of the fjord.

Managing the Environmental Crisis and Restoration

Oslo's long continuous growth slowed down to a halt in 1969, and during the following 15 years 40,000 inhabitants moved out of Oslo. The slowdown was not caused by economic recession, as the Oslo business region flourished and the population in the peripheral local authorities increased dramatically during the same period. It was mainly the young families who moved out. One important reason was a housing shortage; another reason was environmental problems. Young families wanted more space, and preferably a garden too, which was more than they could find in the city where flats were the norm in the new satellite districts. Affordable housing of the desirable kind was available in the suburbs, and the environment was safer and healthier than in the city. The rising wages in the economic boom of the 1960s and 1970s enabled them to afford it. Furthermore, where there was no or little public transport, the car still made commuting possible. The car may have been a convenient solution for the moving families, but its environmental impact was also soon to be the major environmental problem for Oslo.

By the end of the 1960s Oslo's environment was in crisis. The rivers and the fjord were heavily polluted, critically reducing not only the quality of the natural water ecology, but also their potential for recreational use. Waste production rose by 5–7 per cent per year, and the landfills were filled up faster than they could be prepared. The relative improvement in air quality from the phasing out of coal use was overcome by the use of heavy oils in heating, and the rise in fossil fuel driven motorized transport. Both contributed to large and unhealthy emissions of SO_2, NO_2 and dust (PM_{10}), where the impact of dust was aggravated by the emissions from the increasing use of studded car tyres in the winter. Growing motor traffic increased street and road noise levels. Urban development seriously reduced the green structure in the city, and the application of new industrialized forestry methods in the Oslo forest reduced both the biodiversity and the natural and recreational quality. The environmental problems were obvious to all; the public reacted and the alarm bells rang both in the city hall and the national parliament.

After the environmental crisis was generally acknowledged, a fairly stable green lobby both inside and outside the city council has kept the environment relatively high on the political agenda in Oslo up until the present. And although many of the important environmental issues initially were brought forward by pioneers and small pressure groups, often supported by the local media, there has largely been relatively broad and stable political support to prioritize the environment in the consecutive policies and budgets. The common denominator for the environmental engagement from the 1970s onward has been the need and efforts for environmental and urban repair within the built-up area. The main motivation has been, of course, to improve the environment for the benefit of the population itself. Another motivation has been to reverse the flight from the city, and bring taxpayers back. New technology within the environmental field stimulated the adoption of more efficient solutions, and rising wages increased the public's tolerance for rising taxes to pay the bill. Time was overdue for environmental urban repair and renewal. Time was ripe for coordinated environmental management.

Oslo's environmental problems were not unique. They were to a large part shared with other large Norwegian and European cities. The cities had neither the legal means nor the financial resources to tackle the problems; they also lacked an environmental management system with a coherent environmental policy. Demands for action from an awakening opinion stimulated, or forced, the local and national politicians into taking the environmental challenges seriously, and establishing suitable political policies, legal frameworks and administrative competence and capacity. The first national Nature Protection Act was passed in 1970, and a national environmental ministry was established in 1972, one of the first in Europe. The Pollution and Waste Control Act was at last passed in 1981. Environmental considerations were gradually included in the national planning and building code. Environmental departments supervising the local authorities' environmental compliance and performance were established in all county administrations, and in 1992 Prime Minister Gro Harlem Brundtland established a stimulation programme, 'Environmental protection in the municipalities', financing the employment of an environmental 'expert' in each of Norway's 439 local authorities, including the capital. With Norway joining the European Common Economic Area in 1992, the initiative for environmental political and legal development was gradually taken over by the EU, which through its strong environmental commitment produced directives to be adopted by the different national states, including Norway. A central activity of the Ministry of Environment today is to translate the different EU directives on air quality, noise, water, waste, energy and biodiversity and so on, and implement them in the Norwegian legal system.

From Health to Environmental Protection

The city council decided in 1973 that the responsibility for coordination and supervision of the city's environmental protection should be taken care of by the city's health council, which established a separate agency for health-oriented environmental protection. A comprehensive environmental status report was first published in 1977, but no environmental policy was proposed for adoption. In 1986 the political and administrative responsibility for the city's environmental commitment was moved from the Commissioner for Health to the newly established political post of Commissioner for Technical Affairs, later renamed Environment and Transport. The move symbolized a shift in the municipal focus and priority from supervision and control to practical action. The Environment Commissioner had the responsibility for most of the environmental 'sinners', that is, the agencies and companies for transport, sewage, waste, streets and forestry, which also had the competence and capacity to clean up and compensate for damages to the environment. Environment policy strategy also had the responsibility for the more popular environmental tasks like parks, open air recreation and energy efficiency. The environmental health administration remained, however, under the health department, and was finally integrated into the newly established environmental department under the agency for urban environment in 2011.

The first environmental policy strategy, implemented in 1991, not surprisingly placed air quality at the top of the agenda. In 1998, the scope of the city's environmental policy changed the focus from traditional environmental protection, to the broader urban ecology through its new Urban Ecology Programme. As revised in 2002–14, the programme took its value basis in the following statement,

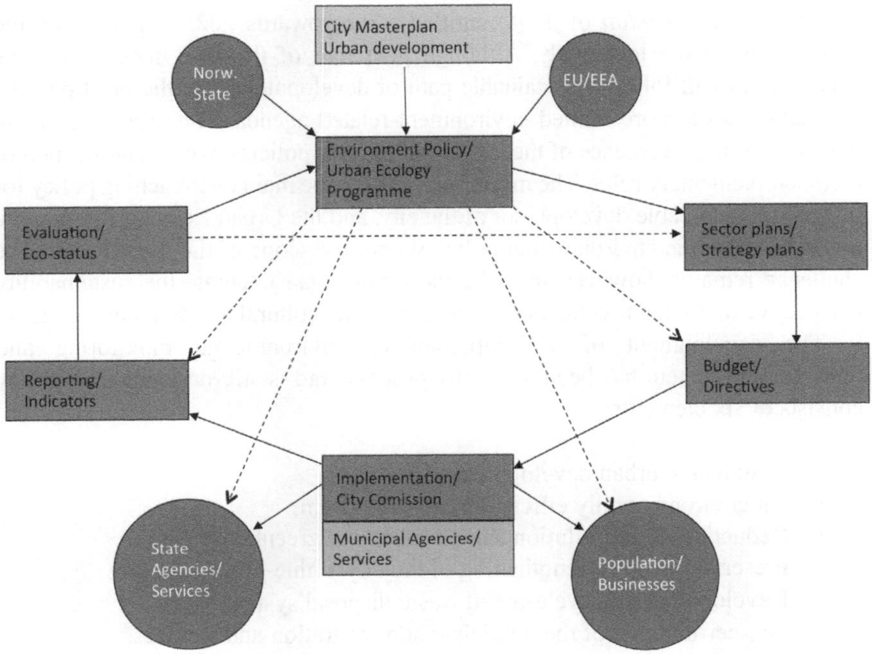

Figure 10.6 Oslo Environment Management System (flow chart)

Source: Department of Environment and Transport, City of Oslo.

'*A city is a culturally created biotope*', and focused on the need to restore the environmental quality of the city according to three main ecological principles:

- Well-functioning ecological cycles for air, water, materials and energy
- Protection and restoration of biodiversity in the city
- The development of a good habitat for the citizens: *Homo Osloensis*

The principles were integrated in the Urban Ecology Programme's six fields of action, with partial goals and indicators for each field of action. For each partial goal was described the strategy/actions necessary to reach the partial goals, including time limit for completion.

Sustainability Transferred to the City Master Plan

Calling the Urban Ecology Programme a Strategy for Sustainable Development promised more than it could deliver, however, since it did not include the social and economic aspects of sustainable development. The environmental department championed instead the integration of the main environmental strategies of the Urban Ecology Programme together with the main social and economic strategies into the new city master plan process, coordinated by the finance department.

The new *Kommuneplan* of 2008, entitled 'Oslo towards 2025', approached the sustainability issue tentatively. Although goal five of the Plan states that 'The City of Oslo will follow a sustainable path of development', for the most part the Plan addresses a more limited environment-related agenda. The process ensured, nonetheless, a convergence of the content of the two policies, with a clarification of their complementary roles. The master plan is now the main overreaching policy for integrated sustainable development of the city, and the Urban Ecology Programme operationalizes the environmental policy within the scope of the master plan. The challenge remains, however, to make the master plan integrate the sustainability perspective of all main sectors, including the social, cultural and economic sectors.

The development of a comprehensive environmental monitoring and management system has been a lengthy process, and is still ongoing. The policy consists of six elements:

- Sustainable urban development.
- An environmentally efficient transport system.
- Reduction of air pollution and emissions of greenhouse gases.
- Preservation and strengthening of the city's blue-green structure.
- Development of a cycle-based waste disposal system.
- Eco-certification of the municipal administration and services.

These policy principles are firmly in place, and the focus now is the need for active and efficient implementation by all parts of the city government.

Urban Development and Transport for a 'Greening Metropolis'

The master plan (Kommuneplan) for 1976–85 focused on developing more 'satellite' urban development in the outer eastern and southern parts of the city, and urban renewal in the inner city.[1] To cover the need to attract young families, the family blocks of flats in the new housing estates were built with three bedroom flats, ample green space for safe play and recreation and local shopping centres. As the building stock in the inner city went into a negative spiral of lack of maintenance and a concentration of welfare clients and, gradually, immigrant occupation, the city established a housing renewal company (Oslo Byfornyelse AS) in 1978. The ensuing ambitious programme turned the negative trend, but came to a premature halt in the economic recession and housing market collapse at the end of the 1980s. Urban renewal was later reinvigorated by a joint national-city government ten-year programme for the upgrading of the Oslo inner east urban environment.

1 The contemporary strategies for, and the development of, Oslo's urban development and transport are described under headings citing major goals in the Urban Ecology Programme of 2002–14.

The main goal of the municipal land use plan of 1991 (and subsequent plans) was to satisfy all needs for new housing and building by densification within the building zone, especially close to the nodes on the public transport system, respecting the limits of the designated 'green border'. To protect the remaining green structure inside the built-up area, a separate plan for the green structure was adopted in 1993, piloting the use of ecological principles in city planning. The policy of urban densification instead of suburban expansion stimulated the search for suitable low-density areas to be redeveloped. An obvious possibility was the port areas, which occupied most of the city's attractive waterfront and barred the public from the fjord. The first project was Aker Brygge, followed by Tyvholmen and Bjørvika. Other major densification projects took place on old industrial quarters in or near the inner city: Akerselva, Nydalen, Løren, Kværner, Ensjø and so on. Existing businesses and the new knowledge-based industries moved up to the large industrial estates in Groruddalen, or beyond the city border. The city and the national government established a joint Groruddalen project in 2006 with a yearly budget of 100 million Norwegian kroner over ten years. Although the improvement of the physical (and aesthetic) environment is still an important part of the project, the main focus is gradually shifting over to the challenges of the multi-cultural transformation and sustainable development of the area.

The environmental quality of the urban redevelopment in Oslo has been rather modest. The new building stock has a relatively high technical and architectural standard, but generally without special environmental qualities or innovations. There are a few exceptions, however. The Aker Brygge mixed redevelopment of an old shipbuilding yard integrates housing, offices, shopping, cafés, restaurants and other leisure activities in a completely pedestrianized environment, including carefully designed and popular public spaces. The urban quality of Aker Brygge has established important standards for the development of rest of the former port areas. The Klosterenga ecological housing quarter completed in 2000 piloted the application of urban ecology principles with an integrated system of energy efficiency, solar power, reuse of local water and recycling of waste (Byggekunst 0304). The redevelopment of the old national hospital near the city centre to the mixed housing, office and educational urban quarter, Pilestredet Park, began with the adoption of an Urban Ecology Programme for the area in 1997. The project was completed in 2008 with a high degree of waste sorting and reuse of building materials, district heating and building energy efficiency, and generous pedestrianized and green areas, including some green roofs (Arkitektur 0308).

An Environmentally Efficient Transport System

The number of motor vehicles registered in Oslo between 1965 and 1985 doubled from 100,000 to 200,000. The increasing environmental and business costs of street and road congestion finally forced the local politicians to seek new solutions, and the master plan for 1980–90 came up with a proposal for a toll fee on the major roads into the city centre. A toll ring company was established by Oslo City Council (County) and the neighbouring county of Akershus to finance 50 per cent of the costs of the new roads for through traffic, including tunnels. The other 50 per cent of the costs were, by agreement, to be financed by the national ministry of transport. Negative public response over time was largely offset by the positive environmental effects. A reduction of 5–10 per cent in car traffic into central Oslo was effectuated, and the construction of tunnels underneath the city further reduced air pollution and noise on the city surface. In addition, 20 per cent of the income from the toll ring was earmarked for public transport. The solution has since been an inspiration for both London and Stockholm in their road pricing schemes. The effect of the 'Oslo road package' was in fact so successful in improving the traffic and environmental conditions in the city that a broad consensus emerged among politicians in the capital region to prolong the toll ring period for another 25 years (Oslo Package 3) and increase the share of the toll ring income earmarked for public transport to 45 per cent, a relatively generous gesture on behalf of the car owners. The logic from a public perspective is quite simple, however: a major city region can never obtain adequate road capacity free from congestion without adequate supply of a supplementary and affordable public transport service. The larger the proportion of commuters who use public transport, the better the access and the lesser the queuing costs for necessary trade, service and other business transport of vital importance for the economic functions of the city and its region.

Slumping passenger numbers on the tram and metro lines led to the merging of all private lines into one municipal public transport company, Oslo Sporveier AS, in 1973. This ensured a better coordination of public transport in the city, necessary to meet the challenge of the steady falling number of passengers in the motor age. The large decrease in the number of tram passengers caused a city council decision to close the tram service in 1960, but the energy crisis in 1973 demonstrated the transport dependency on fossil fuels, and the last four tram lines were saved from closure by a turnaround in the city council decision in 1977. The construction of the eastern metro lines was completed in 1981, and a new metro tunnel underneath the city centre and an outer metro-ring united the eastern and western metro lines into one continuous metro line system. The Oslo Package 2 for the public transport system funded the renewal and increase of the rolling stock and increased the service frequency and upgrading of tram and metro stops and stations. The new transport policy was a considerable success. The decrease in the share of public transport of the total transport in the city of Oslo ground to a halt in 2000, and public transport began to take a larger market share from motor traffic in 2007.

Increasing the transport share of public transport in the larger Oslo region at the expense of car traffic has on the other hand been far more difficult to obtain. In addition to urban sprawl, historic underinvestment in the regional and technical infrastructure of the national rail service has prevented a significant shift in commuter traffic from road to rail. The parliament has, however, accepted its responsibility for providing the necessary rail service capacity to increase the market share, and is committed to doubling the rail track capacity on the three main rail lines into Oslo, and renewing the rolling stock.

Climate Policy

The city's long-term energy (and waste) policy based on hydroelectric power and district heating has contributed decisively to Oslo's very low greenhouse gas emissions. Energy from incinerating residual waste heats 20 per cent of the floor space in the city, and methane from the closed landfill is used to produce electricity. The financing of a municipal energy efficiency fund was initiated in the wake of the world energy crisis in the 1970s, reaching 600 million NOK in 1989, and has since financed energy-efficiency investments, saving the equivalent energy production of a large hydroelectric power station. The city council strengthened the climate policy in 2005 by deciding on the goal of a 50 per cent reduction in greenhouse gas emissions by 2030, and changed the scope of the fund to include phasing out the use of oil in buildings. The reduction of serious air pollution from closing and moving industries, in addition to district heating, was partly counteracted by the dramatic increase in car, and especially heavy, transport. Whilst the pollution of sulphurous acids (SO_2) was gradually reduced, the pollution of dust (PM_{10}) increased, both from exhaust fumes and the increasing use of studded tires in winter. The city decided to introduce a tax on the use of studded tires in the city in the winter season, and the share of cars using studded tires has gradually been reduced to 16 per cent, which, combined with increased street and road maintenance, has brought the pollution of dust down to under the EU maximum levels. The problem remains for nitrogen dioxide (NO_2) and carbon dioxide (CO_2), where the only solution is a major reduction in car traffic, or a shift from the use of fossil fuels to fuels based on renewable energy, or both. The use of electric cars in the city is stimulated by both free passing of the toll ring, free parking and permission to use the bus reserved lanes along the major roads in and out of the city. The number of electric cars in the Oslo region approached 2,000 in 2010, which is a large number compared to most European cities, but is still a very small percentage of all cars in the region.

Despite innovative approaches to reduce the environmental effects of car traffic, road traffic using fossil fuels remains the greatest environmental challenge in Oslo and the outlying region. Norway is a large oil producer, and the influence of the oil industry on the national transport policy should not be underestimated.

Ecological Functioning of the City's Green and Blue Areas

As we have seen, there is a strong interconnection between the vegetation and water resources in the city, both regarding biodiversity and open-air recreation. Biodiversity is nearly always greater at the water edges, where also the experience of nature is often strongest. The protection of the drinking water resources in the city forests bars urban development, and allows open-air recreation.

The struggle to conserve the city's green structure has been waged on two fronts: to preserve the natural and recreational values of the city forest against the use of industrial forestry, and to protect the remaining green areas within the urban region from building and other development. The city forest is part of two large forest areas stretching into the neighbouring municipalities in directions northwest and southeast of the city. The forests are of substantial economic significance, with many landowners practicing industrial forestry as the city's own forestry agency did for many years. Pressure from environmental NGOs has resulted in a change of policy for the management of the city-owned forest, with the adoption of the principle of multiple use and balanced resource management, where the considerations to recreation and nature research should override the profit motive. Pressure from the local branch of Friends of the Earth stimulated the designation of natural reserves for the protection of important biotopes in parts of the forest less effected by forest exploitation. The process culminated in the designation in 2009 of the whole forest beyond the 'green line' (1,700m^2) to a natural park of national interest, balancing the interests of nature protection, outdoor recreation and sustainable forestry.

The goal of protection and strengthening of the green structure inside the city urban area, however, has proved to be a tough challenge to meet. A large part of the green structure in the city is private, mostly gardens and green areas on industrial and commercial sites. A policy of densification represents a serious threat to the protection of open green space. A green land use plan for the urban area was adopted in 1993, based on the principles developed by the city's first urban green landscape ecology analysis. This was supplemented by a set of 'green posters' with information on the different urban districts' administrations of the green values to be protected or considered in their districts. The biodiversity of the green areas in the built-up area was later registered and digitalized for easy use in planning and building permit handling. The city's practical strategy has been to allow densification at the expense of green areas on private land designated for urban development, and to concentrate the efforts on the protection and/or acquisition of green areas specifically designated as part of the city's green space. The public leniency towards the strong pressure to use green space for building much needed kindergartens and schools has gradually been reversed, and the total area of legally protected open space in the city has actually increased during recent years. The main challenge today is the wear and tear on the existing parks and green areas from a growing population, and the struggle to maintain, and improve, the ecological, aesthetic and functional quality of the green areas. The time had come to regard the potential of green roofs and walls of the building stock as an important part of the future green structure of the city, and the first modern building with a green roof appeared in Pilestredet Park in 2006.

The efforts to restore the water ecology of the rivers and the inner fjord started with the construction of the large new sewage treatment plants Bekkelaget in

1963 (rebuilt and modernized in 2001), and VEAS (west) in 1982, and continued with the completion of the sewage duct network system connecting all pipes and ducts to the treatment plants in 1987. Leakages and overflows of polluted urban surface water into the rivers and the fjord continued, nonetheless, especially during periods of heavy rains. The city has launched a large sewage renewal project 'Midtgardsormen' to lead overflow to Bekkelaget, enabling swimming in the inner harbour from the waterfront of the new Bjørvika development. Another ecological challenge has been the protection of the city's rivers and brooks, and a set of 'Environmental policy principles for protection of rivers, brooks and lakes' was adopted in 1982, stating the goal that 'all rivers and streams should be in ecological balance with a natural biodiversity'. This prevented the encasing of rivers in built-up areas, and secured a building-free zone of 20 metres on each side of the river. The first reopening of a ducted river took place on a stretch of the Alna River; 'Hølaløkka', in Groruddalen in 2005. A major force in championing the environmental and educational value of Oslo's rivers has been the Oslo River Forum, an association of mainly public-spirited pensioners, established in 2000.

Waste Disposal

The steadily increasing number of household and business waste forced Oslo to look for alternatives to increasingly unpopular landfills. The programme committee for technical affairs stated in 1972 that 'the long term goal for the handling of the problems of pollution and resource management regarding sewage and waste must be the future preservation of natural resources by extensive use of recycling'. Oslo's first waste recycling plant REO opened on *Klemetsrud* in 1985 after a long process of planning and doubt. The process was meant to recycle 60 per cent of all paper, 50 per cent of all plastic and 85 per cent of all magnetic iron. Residual waste was to be incinerated and the energy used for district heating. The time was ripe, but the technology was not, nor was the materials market. One of Europe's first comprehensive waste recycling plants for treating curb-side collected waste was forced to close in 1988, after considerable loss of both money and prestige for the city and its private business partners. It will probably take another 20 years before the city ventures a second major attempt on the same scale. Since 1988, waste handling practices have been based on curb side collection of paper and residual waste, whilst households were encouraged to sort glass, metal and hazardous waste to be delivered at designated return and reuse stations. Residual waste has been incinerated and the energy used in the district heating.

The city decided finally in 2005 to make another attempt at a cycle-based waste system, based on the sorting and collection of plastic and kitchen or wet-organic waste. An optic sorting plant was opened in 2010, and a new biogas production plant outside the city will produce methane and garden soil from the wet-organic waste. The processed methane from waste and waste water sludge will be able to fuel approximately 250 buses in the region. To reduce the amount of waste, and increase recycling, the city cooperated with the NGO Green Living to recruit 'green families' and stimulate households to adopt eco-friendly lifestyles with reduced material consumption and preference for eco-labelled goods and services. The annual *Rusken* campaign has, since 1976, mobilized over tens of thousands of small and adult inhabitants to clean up communal lawns and lanes for paper and other rubbish and waste when the snow disappears in the spring. On the other hand, the amount of tagging and other 'graffiti' on private and public buildings and installations has increased over the last few years at considerable costs for both the city and the community.

Conclusion

The general pattern of Oslo's environmental development has many parallels to that of most European capitals. Norway has been under strong cultural influence of impulses from northern Europe, and especially other Scandinavian countries. In recent years, as member of the European Economic Area, Norway has had to comply with EU legislation concerning the environment, reinforcing the European influence on Norwegian environmental priorities and solutions. Oslo's participation in the European Sustainable City and Green Capital Award contests is a significant symbol of this, and the results from this benchmarking with other leading European 'green' cities have inspired the revision of its environmental policy, as adopted by the city council in 2011. There are, however, certain features about Oslo's environmental history, present state and future challenges that, though not entirely unique, are singular and therefore worthy of consideration.

One typical feature is the strong role of nature protection in the environmental history and current state of Oslo, reflecting the Norwegians love of nature. For many citizens, environmental protection has been synonymous with nature protection and the main symbol and object is the *Marka*. This large boreal city forest surrounding the city to the west, north and east, represents two-thirds of the area within Oslo's municipal borders. The struggle has been successful. *Marka* is now protected by the strongest possible legal means available: a special *Marka* law, passed by the national parliament, protects the city forest against urbanization for future generations, and regulates the multiple uses and balanced management of the forest. Being aware of this nearby and being an extensive natural area may have made Oslo citizens and politicians more tolerant of plans to densify the built-up urban area.

Oslo's steady efforts over the years to reduce the impact of environmental challenges, and repair environmental damages, must also be characterized as fairly successful, although there is still important work to be done regarding the protection of green areas and biodiversity in the city and further reductions in the pollution of air and water, cleansing of polluted soils and so on. Oslo has lower climate carbon emissions per person than most European cities, and the challenges of climate mitigation and adaptation are on top of the city's political agenda. These global challenges cannot be solved, however, by the application of just more of the same traditional environment protection methods that have worked thus far. The necessary fight against both direct and indirect climate gas emissions requires more than only more plans and resources: it requires a shift in paradigm. It affects both our lifestyle and our consumption pattern. Environmental protection thinking has to be replaced by ecological thinking, and sustainable ecological cycles made the new norm.

Oslo has seemingly taken this step, in calling its environmental policy an 'Urban Ecology Programme'. Since traditional environmental protection thinking is so deeply rooted, both in and outside the public bureaucracy, the effect has so far been more symbolic than factual, but it represents a start. Another factor has been the preoccupation of the city urban planning and building sector with high-tech aesthetic or 'land mark' architectural quality rather than eco-architecture with innovative environmental quality. This is especially demonstrated in the prestigious urban waterfront development of Bjørvika, where the innovative ecological quality of the new public and private buildings so far lags behind comparable waterfront developments in other 'green' cities like Stockholm, Hamburg and Malmø (Sweden).

One hope for the future lies in Oslo's emerging Green Wave, where the eco-certification of the municipal administration and services not only increases the eco-efficiency, but also stimulates the environmental consciousness and motivation of 40,000 employees, as well as the children and parents of Oslo's 500 schools and kindergartens. By practicing eco-procura, the city creates a profitable business market for eco-efficient goods and services, and stimulates the eco-certification and eco-efficiency of a growing number of private businesses. The effect of the Green Wave has so far been limited by the national government's reluctance to eco-certify the large state sector in the capital. The necessary paradigm shift in climate and eco-efficient behaviour requires leadership, and there are limits to the effect of the city's positive example without the active support of the national government.

Bibiography

Arkitektkontoret GASA. 2004. *Klosterenga* økologiboliger. Byggekunst 0304.

Blågrønn Hovedstad; jubel og gråt for Oslos elver og bekker. 2010. Oslo Elveforum.

Butters, C. 2008. *Pilestredet Park; et* økologisk *tilbakeblikk*. Arkitektur 0308.

Dagre, T. (ed.) 1980. *Natur og miljø i Oslo; Status -77*. Universitetsforlaget.

Economist Intelligence Unit. 2009. *European Green City Index; Assessing the Environmental Impact of Europe's Major Cities*. Siemens.

European Sustainable City Award 2003; Guidelines, Oslo's Application, and the Jury's Verdict. European Sustainable Cities and Towns Campaign. Brussels. 2003.

Fossum T.I., Tveito, D. 1991. *Den Grønne By; Oslo park og idrettsvesen gjennom 75 år*. Grøndahl & Søn Forlag AS.

Grundt, G. 2000. *Hovedstadens miljøutfordringer i tusenårsperspektiv*. Kommunalteknikk 11/12.

Grundt, G. 2007. *Er Oslo bærekraftig?* Plan 3/4.

Johansen, T.A. 2001. *Under byens gater; Oslos vann- og avløpshistorie*. Vann- og avløpsetaten, Oslo kommune.

Lund, B.H. 2000. *Beretning om Oslo kommune for* årene *1948–1986*. Oslo kommune.

Measuring Urban Sustainability; Analysis of the European Green Capital Award 2010 & 2011 application round. European Commission/Ambiente Italia. 2010.

Miljøstatus – 92; Natur og miljø i Oslo kommune. Etat for miljørettet helse; Oslo kommune. 1993.

Miljøstatus for Oslo. Friluftsetaten; Oslo kommune. 2010.

Oslo; Planning and development; A Survey of the Geographical and Historical Background, The Development of Population and Economic Activities and Post-war Planning and Building. City planning department; City of Oslo. 1960.

Oslo towards 2025; The Municipal Master Plan. City of Oslo. 2008.

Oslohelsa; Utredningen om helse, miljø og sosial ulikhet I bydelene. Klinikk for forebyggende medisin; Ullevål sykehus; Oslo kommune. 1997.

Oslomiljøet; Byøkologisk program. Byrådsavdeling for miljø og samferdsel; Oslo kommune 1998.

Rullering av Byøkologisk program for 2011– 2027. Oslo kommune, 2011.

Sten, S. 1945. *Oslo gjennom tusen* år. John Griegs Forlag.

Stor-Oslo; Forslag til Generalplan. Det Mallingske bogtrykkeri. Oslo. 1934.

Strategy for Sustainable Development; Environmental and Sustainability Status 2002 and Urban Ecology Programme 2002 – 2014. Department of Environment and Transport; City of Oslo, 2003.

The Expert Panel's Evaluation Work & Final Recommendations for the European Green Capital Award of 2010 and 2011. European Green Capital, 2009.

Torstenson, I. 1997. *Fra Nattmann til renholdsverk; Avfall og renovasjon i Oslo gjennom tusen* år. ProArk AS.

Towards a Thematic Strategy on the Urban Environment. Commission of the European Communities. Brussels, 2004.

Urban Development in Oslo; A Brief History of Oslo and Urban Development in the 90's. Agency for planning and building. City of Oslo, 1994.

Urban Sustainability – Learning from the Best; Catalogue of Best Practice; European Green Capital Award 2010 & 2011. European Commissioner for the Environment/COWI, 2010.

Chapter 11
Planning for a Green Oslo

Karsten Jørgensen and Kine Halvorsen Thorén

Introduction

Green urban planning has attracted growing attention over the past ten to twenty years (Fabos 1995, Fabos and Ryan 2006). One important reason for this – at least from a Norwegian perspective – is the densification policy that was introduced in the 1990s and that put additional strain on green spaces inside town and city boundaries. Up until then, urban development had for the most part come about as a result of urban expansion, and the urban growth that has taken place since World War II has been extensive in terms of land use. The publication of the Brundtland Commission report 'Our Common Future' (World Commission on Environment and Development 1987) led to urban development in Norway being perceived as unsustainable (Næss 1992) due to poor utilization of land resources and the extent of urban sprawl across the landscape. Arguments in favour of more densely populated urban development were linked to reducing CO_2 emissions and private car use, protection of cultivated land, conserving biological resources in both forest and agricultural production, safeguarding biodiversity and recreational activities in undeveloped rural areas and so on. A report to the Storting (the Norwegian Parliament) in 1993 therefore ascertained that Norwegian urban development should be densified (Miljøverndepartementet 1993), a decision that was not without controversy. One of the reasons for this being that densification may affect green spaces, something which was also acknowledged in the parliamentary report, in which the need to prioritize and plan the conservation of green spaces was emphasized: 'green' urban planning therefore regained favour, something that had not been the case for some years.

The growing interest in reviving 'greenness' as an element of urban planning also aroused our curiosity and serves as the background for this chapter. What scientific and political decisions – or even ideologies – lie behind the pattern of green spaces that has emerged today? Are there lessons to be learned from previous green urban planning projects that can help us understand today's situation? We have chosen to take a closer look at Oslo's recent planning history in order to find out more. Oslo is proud of its traditions in the planning of green and natural spaces, even if – as we shall see – this has varied from one period to another. The city has also undergone large-scale migration in recent years, and the strain on these areas is considerable. This has resulted in widespread popular mobilization in defence of green spaces, as reflected in the establishment of the grassroots movement Grønne lunger Oslo (Green Lungs for Oslo). This engagement has also led to increased interest by both politicians and urban planners.

Knowledge Gaps

Much has been written about Norwegian urban planning history (see for example Jensen 1980, Tvinnereim 1990) and urban planning of Oslo in particular. When it comes to Oslo, this covers everything from general outlines to specific periods, plans or topics. For example, the publication entitled 'Oslo. Planning and Development' (Byplankontoret Oslo kommune 1960) provides a broad outline of Oslo's development up until the 1960s. The first period in the modern history of urban planning in Oslo to have attracted great interest began with the publication of Harald Hals's plans in 1929 and extended to 1934 (Lexau 1993, Aspen 2010). Much has been written about the geographical development within the urban districts, such as *Kvadraturen* (Berg 1965) and the new satellite towns and so on. Housing policy and housing planning (Hansen and Guttu 2000) are other topics that have been covered.

Oslo has also received some attention from a green perspective, though this is often in relation to the expansive green belt (*Marka*) surrounding the city (Moland 2006), or the parks and public green spaces or covers the early history (Bruun 1984; Bruun 1992). The topic which has attracted most attention is park policy, particularly in connection with the history of the Parks Department. This history has been told twice, first by Marius Røhne, Oslo's first municipal gardener (Røhne 1967), and later by Ida Fossum Tønnessen and Dagfinn Tveito (Tønnesen and Tveito 1991). Little of Oslo's green history has been published over the past 20 years, and surprisingly little about the connection between urban planning and planning for what today is usually referred to as the city's 'blue-green structure'. It is therefore our view that it would be interesting to take a more systematic look at Oslo's green urban planning history from the 1900s up to the present day.

Research Question and Method

The main theme of this chapter is: how, and to what extent, has the green structure been integrated into the urban planning of Oslo, and what significance has been ascribed to 'greenness' in key planning documents from the 1900s through to the present day? This is an ambitious project, so our intention is therefore simply to start a process which hopefully will lead to more in-depth studies of the subject.

We use the term 'green structure' in line with what is usual in key Norwegian documents on the subject such as guidelines from the authorities, reports from the Norwegian parliament and so on (Miljøverndepartementet 1993, Thorén and Nyhuus 1994). In these documents, green structures are defined as all areas in the city that are covered by vegetation, irrespective of whether or not they are located on private or public land. Also included in this definition are natural and cultivated vegetation, lakes, rivers, streams and so on. Because of this, we often refer to the 'blue-green structure', a term of which municipal authorities are particularly fond, as is apparent in the city's slogan, 'The blue and the green and the city in between'.

For purposes of presentation and analysis, we have chosen to group the material into five main periods. The criterion for this division is based on our interpretation of when important changes in the perception of urban planning or urban development influenced 'green values'. This resulted in the following chronology:

1916: Emergence of public parks in Oslo.
1916–1948: The parks system approach is introduced.
1948–1980: The green structure grows and culminates.
1980–2009: Discontinuation and decline.
2010: Today's green structure.

We also had to delimit the subject thematically, and have therefore completely omitted the planning processes and the comprehensive material relating to securing the green belt and the establishment of the Act of Parliament for these areas, which came into force in 2009. Nor will we discuss the municipal Parks Department's role in great detail, due to its own particular significance and scope.

We have concentrated on studying the master plans prepared by the City of Oslo, starting with the first municipal master plan adopted in 1929 (Hals 1929) up until the present day. To shed some light on the period prior to 1929 we used Marius Røhne's detailed account entitled 'Oslo kommunale parker og grønnanlegg 1810–1948' ('Oslo Municipal Parks and Green Installations 1810–1948') (Røhne 1967) and the results of recent research on this subject in Oslo, such as *Norsk hagekunst under funksjonalismen* (*Norwegian Landscaping during the Functionalist Period*) (Apall-Olsen 2007). The following types of plans are included in the review: a few regional plans, general plans,[1] municipal plans and thematic municipal sector plans which are significant to the subject, particularly those dealing with the city's green spaces.

The literature was reviewed with the aim of identifying what had been written about the full spectrum of green spaces, on both private and public land, in line with the definition of 'green structure' given above. Briefly put, this means that we have structured the discussion of the different periods around the following main themes:

- Important plans: growth and growth directions;
- Intentions and realities of municipal green structure planning;
- Influence and ideology.

In the next section we will present the periods, and in the final section we will discuss to what extent green structures have been integrated into urban planning

1 Up until 1984, in connection with amended legislation, when the plan type began to be referred to as 'municipal master plans'

in Oslo, and what significance has been ascribed to greenness in key planning documents from the 1900s and up to the present day.

Green Spaces in Oslo before 1916: Emergence of Public Parks in Oslo

The first period of our chronology is marked by a fragmented and random development of green spaces. There were a number of smaller, publicly accessible green spaces established through private initiatives. The city's residents' association (*Selskabet for Christiania Byes Vel*), established in 1811, held the building of new parks as their main objective during the first 50 years of its existence. The society instigated 'The Promenade' in 1812, opened to the public in 1819 (the remains of which is today known as Børsparken) in addition to a number of other small parks in the following years. The central mall in modern-day Oslo, Eidsvolls plass, was purchased and left as green spaces by landowners on the opposite side of the newly established main street, Karl Johans gate, between the city and the Royal Palace in 1846, to avoid development which could obstruct the view of the Oslo fjord. In 1858 the government bought these areas in order to develop a public park adjacent to the area known as Studenterlunden, established as a green space on erection of the university buildings around 1840.

King Carl Johan supported the establishment of green spaces in Oslo soon after he inherited the Swedish-Norwegian throne in 1818. He bought properties belonging to the former Royal Manor at Bygdøy between 1819 and 1837 to build a public promenade, and initiated the building of the Royal Palace closer to the city. A major park was planned around the Palace and built around 1840, and the idea was that there should be a contiguous series of parks connecting the Palace Park and the public promenade and woodland park at Bygdøy. This plan was not realized, but the Royal Palace Park became one of the major public parks in central Oslo and Carl Johan and his successors developed Bygdøy, officially announced in 1837, into the city's major green space. Both parks offered public walkways, viewing points and sculptures and, at Bygdøy, restaurants (Jørgensen and Fredriksen 2008).

The city authorities did not regard public parks as part of their responsibility to begin with. In 1857 a sum of money was granted for the development of the St. Hanshaugen Park, a park that was initiated by a private citizen and supported by the city's residents' association. In 1860, the municipality accepted responsibility for the further development and maintenance of the park.

During the last five decades of the 19th century, the population grew fivefold, from 50,000 to 250,000, and the city council slowly began to take on more responsibility (Byplankontoret Oslo kommune 1960). In 1875 the Planting Department was established, and from then on the city regularly supported both the establishment and the maintenance of parks. But there was still no cohesive park planning policy. The parks that were built between 1860 and 1990 in the new and densely populated working-class neighbourhoods on the east side of the

Figure 11.1 The map shows green areas in Christiania 1858. Few of these areas were available for the city's population at this time

Source: Rolfsen 1950: 25.

city, such as Grünerløkka, were realized thanks to single private developers. This resulted in well-planned areas of parks covering the size of one whole block, as in Paris, as an integrated part of the scheme.

The City of Oslo and its Planting Department did realize some street planting projects. Bygdøy Allé, built in the 1890s and planted with horse chestnut trees, is a well-known example. The City of Oslo also purchased special areas to secure access to green spaces for the general public. Another important example is the acquisition in 1881 of the hillside at Ekebergåsen. These improvements certainly enhanced the city's green image, and in addition improved access to green spaces for the inhabitants of the east side of the city. Both examples show the change in attitude on the part of the city authorities that occurred around 1870. In general, however, the period was characterized by scattered and private developments, with no well-developed park planning philosophy, nor with strong political arguments for the parks.[2]

The arguments used by the city's residents' association were mainly aesthetic: there should be parks in the city to enhance its beauty. For developers, and to an increasing extent for the politicians who engaged in this line of reasoning, developing the parks compensated for the poor living conditions that were offered in these densely populated areas. Another reason was to contribute to aesthetic pleasure and proximity to nature in order to enhance the physical and mental well-being of the inhabitants and also strengthen public morale. This is in accordance with the recommendations by C.C.L. Hirschfield, a Danish philosopher and author of *Theorie der Gartenkunst* (*Theory of Garden Art*) (1779–85).

For the king, an important reason was the nation-building process and the city's reputation; Carl Johan wanted the Norwegian capital to be representative and comparable to the Swedish capital, Stockholm, and to other European cities. Carl Johan had been a general in Napoleon Bonaparte's army and the French deputy to Vienna, so he knew many examples from European cities where parks were to an increasing extent becoming an integrated part of urban development.

1916–1948: the Parks System Approach is Introduced

The very rapid growth in population during the latter half of the 19th century slowed down after 1900, but the number of inhabitants continued to increase, and the development of residential areas was a major issue. The first decade of the 20th century did not see many projects realized, but between 1916 and 1934 a large number of publicly built or supported residential areas materialized. More than 10,000 housing units were built, and the standards for green spaces in relation to the residential areas increased significantly during that same period. The main developments were Tøyen, Lille Tøyen Hageby, Ullevål Hageby, Lindern, Torshov, Rosenhoff, Vøyenvolden, Jessenløkka and Ila, and many of them were based on garden city principles.

We will highlight two master plans that had particular significance for Oslo's green structures during this period. One of them is Marius Røhne's[3] plan for the development of Oslo's parks and green spaces from 1917–18 'Parker og grønnområder i byen' or 'Parks and Green Areas in the City' (Bruun 1992: 153).

2 Thomas Hall observes: 'An examination of a late nineteenth-century map of Christiania [...] reveals remarkably little of the typical urban planning features of those times; there is hardly a single tree-planted street worthy of the name of boulevard or avenue, few central park areas apart from Studenterlunden, little in the way of rectilinear elements and only one or two attempts to create monumental square' (Hall 1997 in Aspen 2003).

3 Marius Røhne established one of the earliest landscape architecture offices in Norway in 1913, and was appointed director of the Parks Department in 1916, a position he held until 1948.

The other is Harald Hals's legendary plan entitled *Fra Christiania til Stor-Oslo* (*From Christiania to Greater Oslo*) (1929) and its sequel (*Generalplankomitéen for Stor-Oslo* 1934).

In 1916 the Parks Department was established, and what had previously been fragmented and arbitrary development gradually took on a more systematic approach. The department inherited a few projects of great importance, both for the general public and for the standing of the new department, such as Frogner Park, Ekebergåsen and the parks on the east side of the city. These were further developed by the Parks Department under Røhne's leadership.

Just outside the densely developed areas to the west of Oslo, the first sections of what would become Frogner Park opened in 1904 as a public park; the so-called 'Gamlehagen' (The Old Garden) which was part of the historical garden that once belonged to Frogner Manor. In preparation for the exhibition in 1914 to celebrate the centenary of the Norwegian Constitution, other areas on this property, which had belonged to the city since 1896, were developed into park spaces. The exhibition grounds from 1914 were later developed into today's Vigeland Park by the sculptor Gustav Vigeland and landscape architect Iosef Oscar Nickelsen.

The development of park areas in the eastern parts of the city, along the Akerselva River, became part of a new planning focus, where green spaces for the general public in and around the residential areas were given priority over prestigious park developments in the city centre. One such example is Grünerløkka, where green spaces were integrated and developed into proper parks and linked by the Akerselva riverside project, starting with the areas close to Nybrua and Ankerbrua. Birkelunden was completed in 1917, and more parks in the same area soon followed, with Torshovparken and Torshovsdalen as the major areas in the 1920s and 1930s. The Akerselva idea was to create a contiguous green space through the east side of the city, from the city centre to the periphery along the river. Due to complicated ownership structures, the work was difficult and marked by many delays. Another important example from the period is Ekebergåsen; a major achievement for the greening of Oslo. Ekebergåsen was also developed as a combined residential and urban recreational area, with large grass fields for sport. Røhne continued to be engaged in this project throughout his career as parks director, and the park was completed at the time of his retirement in 1948.

This focus on parks and green spaces including housing quality was the joint effort of Marius Røhne and Harald Hals, and the municipal plan from 1929 shows that Hals shared many of the ideas in Røhne's plan from 1917–18 (Aspen 2003: 234). As planning director, Hals contributed significantly to the modernization of Oslo, including the addition of green spaces in the master plan, as part of the planning philosophy. In the municipal master plan from 1929 Harald Hals says: 'It is *our time* that has made parks and green spaces a very important and absolutely indispensable part of the urban organism' (Hals 1929: 182). Although Hals and Røhne did not seem to have cooperated directly with one another, they were in most cases working towards the same goals. It was less a question of defining

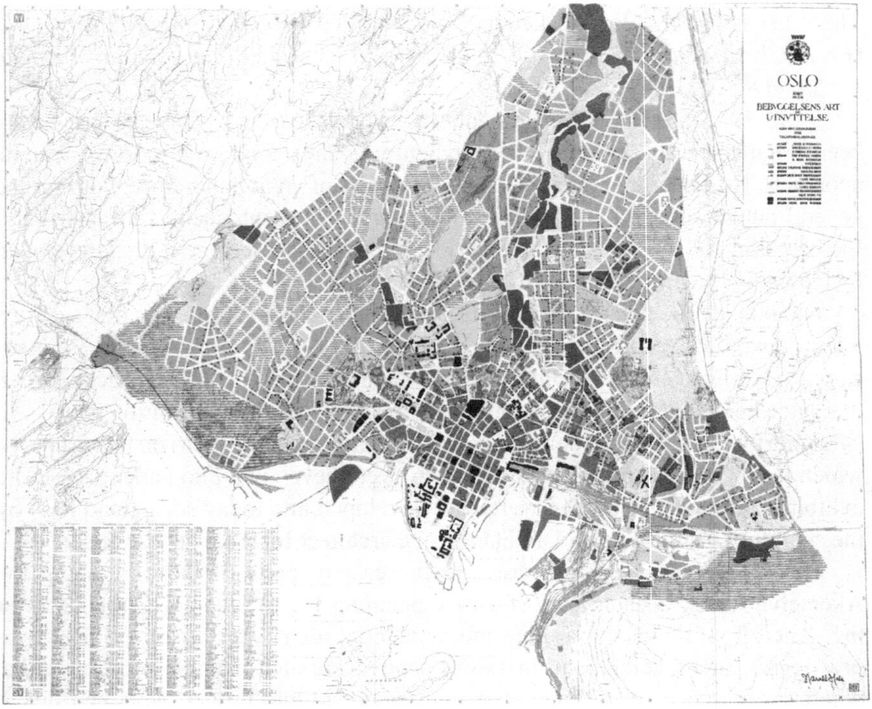

Figure 11.2 The park system as an integrated part of the municipality plan

Source: Harald Hals (1929: Fig 87).

standards for green spaces in residential areas and more an attempt to emphasize the virtues and advantages of having green spaces blended into the urban structure.

It is important to note how the forests and parks and so on appeared as main elements of the municipal master plan, and also how Hals used 'modern' methods to estimate the quantity of green. This relates well to his statement regarding the parks system; the idea was not only to satisfy standard requirements for playgrounds and sports facilities, but also to 'force all these isolated areas together in a cohesive parks system' (Hals 1929). In the initial mission statement for the newly created Parks Department in Oslo 1917, the idea of a parks system was launched (Bruun 1992: 153). According to Røhne, the parks should be integrated into a system of radial arteries, or 'green fingers'. This was obviously a better and more effective basis for a green park policy. The parks system included urban greenness of all kinds, but also the great forest areas that surrounded the city. Nature areas were seen as an important part of the air purification system, so improving the air quality was also part of the Hals/Røhne vision for Oslo's park system.

These development's during the first decades of the 20th century partly reflected Oslo's transformation into a major European capital, with considerable self-confidence. The dissolution of the union with Sweden in 1905 meant full independence, a sovereign king and an independent government. The need for more and better residential areas was a main issue, as was the question of acquiring sufficient building land. At the beginning of the 1900s extensive suburbanization of the neighbouring municipality, Aker, began. The population in Aker had grown more than fivefold between 1900 and 1946 (Utviklings- og kompetanseetaten and Satistics Norway 2009). This was made possible because of extensive development of the road network. Thus, most of the available area was developed, and Oslo needed more space to develop further. The boundary adjustments between Oslo and Aker had not kept pace with developments in population growth. Modification of the boundary was no longer an option, and in 1948 Aker and Oslo therefore merged.

It is significant to note that there was a strong focus on the need for green spaces in the city in general, especially in the eastern parts of the city, and in relation to the residential areas. Historically the eastern parts of the city are developed in line with the industry as working-class residential areas, and thus with higher density and fewer green spaces than the western parts of the city. The ideas for developing parks systems were imported from European and American urban development projects, and the arguments were mainly related to public health. Examples from America seem to have been particularly important in relation to the development of public green spaces.[4] Hals studied at the University of Illinois between 1902 and 1906, and most likely visited the famous residential area of Riverside, Illinois and other projects by Frederick Law Olmsted, such as the parks and parks systems in Boston, Buffalo and New York. Hals was also clearly influenced by the garden city philosophy of Ebenezer Howard (Howard 1902).

1948–1980: The Green Structure Grows and Matures

It was not until 1950 that Harald Hals's plan for Oslo was replaced by a new master plan (Rolfsen 1950). The merger of Oslo and Aker municipalities in 1948 made it necessary to update the plans, which were used longer than intended due to the World War II (1940–1945) (Rolfsen 1950).

In the municipal master plan of 1950 the focus was placed on the outer zone, that is, the urbanized section of the Aker municipality, which was in dire need of systematic management. The main objective for the municipal master plan of 1950 was therefore to prepare new development areas in Aker to provide enough housing for the population and, not least, better growth opportunities for industry, which was squeezed into Oslo's inner city district. In the next plan, in 1960, the

4 The 1914 volume of the leading Norwegian art journal *Kunst og Kultur* is a special issue on garden art. There are two articles about urban development and public parks in America; one by landscape architect I.O. Nickelsen, 'Byer med trær, lekepladser og blomster' ('Cities with Trees, Playgrounds and Flowers') and the other was professor of history of art Harry Fett's *Amerikanske lekeplasser* (*American Playgrounds*), clearly indicating that America was the most interesting place in this field.

focus reverted to the city centre and to Oslo as it had been originally conceived. This plan also abandoned Harald Hals's principle of concentric urban growth (Byplankontoret Oslo kommune 1960). There was no longer a wish to continue as before with open, detached housing developments on the outskirts of the city. Instead of decreasing density from the centre outwards, the idea was to have a series of secondary centres and sub-centres connected by green corridors and systems of nature areas and parks.

The focus on the inner city district and the living conditions there prevailed in the next plan, which came in 1975 (Oslo 1975). Urban sprawl and the release of private passenger vehicles for sale exacerbated housing conditions in central Oslo, which were already sub-standard. This situation resulted in a decline in the population. Improving housing conditions in order to retain families with children was therefore a key element in the municipal master plan of 1975. Urban renewal of the inner city was given priority, and the objective was to create larger apartments and better outside areas with more sun, light and air, that is, reduced plot utilization.

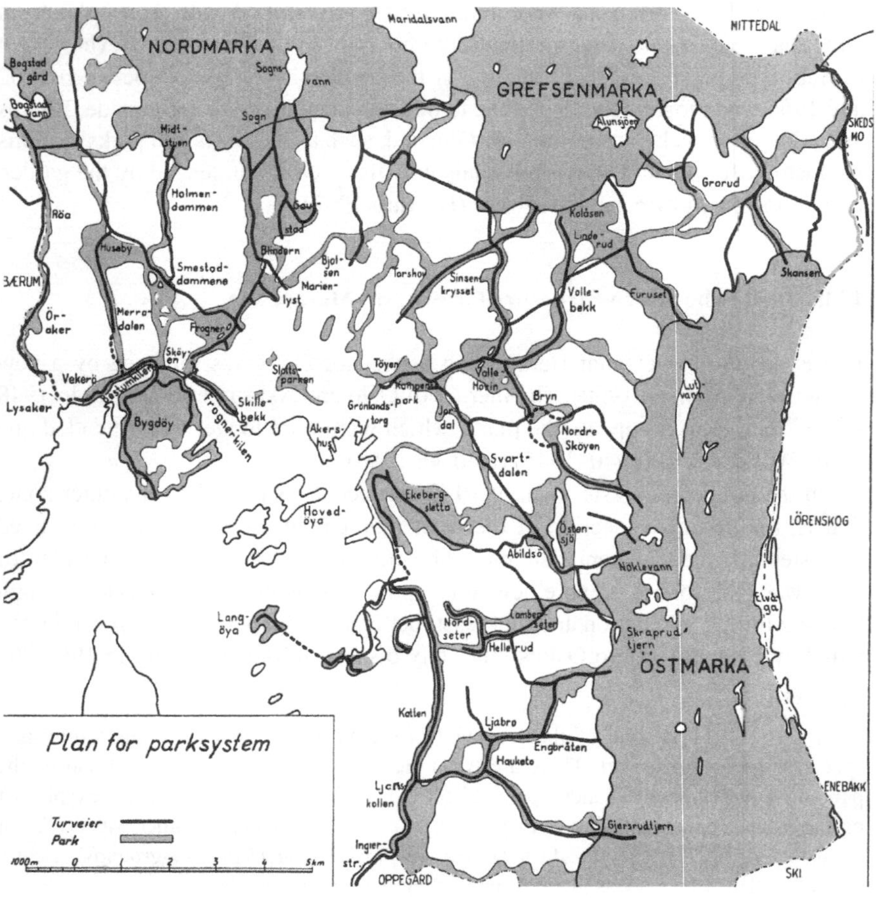

Figure 11.3 Plan for the parks system and trails in Oslo (1949) shows how the green spaces, rivers and landscapes would divide and link the city; the same principle is repeated at the neighbourhood level in Lambertseter

Source: Byplankontoret, Oslo kommune 1960: 38, 42.

The improved standard should contribute to maintaining the population but not necessarily to increasing it. 'Oslo is for the most part a developed municipality', according to the municipal master plan from 1975 (Oslo 1975: 76).

Between 1969 and 1983 the population of Oslo fell by as much as 40,000 (Statistics Norway 2011), a situation which gradually caused concern because it affected tax revenues and, consequently, municipal finances. The municipal master plan of 1980 therefore represented the beginning of a new phase that has proven to be of major significance for the city's green spaces. Increased housing production was highlighted as one of the few municipal measures for raising revenues (Oslo kommune 1980). Nonetheless, the plan emphasized that future development should not deprive those who already bore the heaviest environmental impact of access to green areas (Oslo kommune 1980). Therefore, different alternatives for urban development were discussed which included building in Oslo's green belt. But insofar as this area had remained an undeveloped reserve, as noted in by Hals's plans, there was a great reluctance to begin development there.

The Parks System and Trail Plan of 1949 is an important basis for the municipal plans of 1950, 1960 and 1975 (Rolfsen 1950, Byplankontoret Oslo kommune 1960, Oslo 1975). The plan entailed a continuation of Harald Hals's concept of a parks system. The idea was that the suburbs should be 'separated from each other by wide green belts, park areas, open stretches, woodland areas, etc. Inside each "town", housing developments would be grouped into neighbourhoods which in turn would be separated by smaller green spaces which linked the neighbourhoods together and which would also be connected to the entire trails and parks system'. (Byplankontoret Oslo kommune 1960). The entire development pattern was based on the principle that the development area should be surrounded 'by Oslo's forest and the city's residential area connected to the forest by a system of green belts and park areas'. (Byplankontoret Oslo kommune 1960)

According to Rolfsen, 'A well planned trail is a nerve system in the city's parks scheme in that it connects the parks with the recreation areas outside the city. In addition, the more a trail can connect local centres, schools, sports grounds, etc, the better' (Byplankontoret Oslo kommune 1960). It was also important to exploit the resources which a landscape had to offer. Rolfsen therefore emphasized that 'a parks plan for Oslo cannot strictly adhere to the old familiar system of park strips and park rings; it must be based on the landscape's natural features, rivers, ridges, woodlands and perhaps to a greater extent the existing open areas'(1950: 62).

The merging of Oslo and Aker was essential to the realization of an aggressive green policy such as that which the City of Oslo had in mind. The reality was that the 'original' Oslo had no more land left to use for green purposes. The merger therefore offered the opportunity to acquire new green spaces and to link green spaces in built-up areas together with the green belt around the city. Aggressive municipal land acquisitions in Aker to secure land for residential development also provided good conditions for planning the trail system. Furthermore, the municipality took advantage of whatever opportunities arose to secure property in the green belt surrounding the city. The Outdoor Recreation Act of 1957 provided

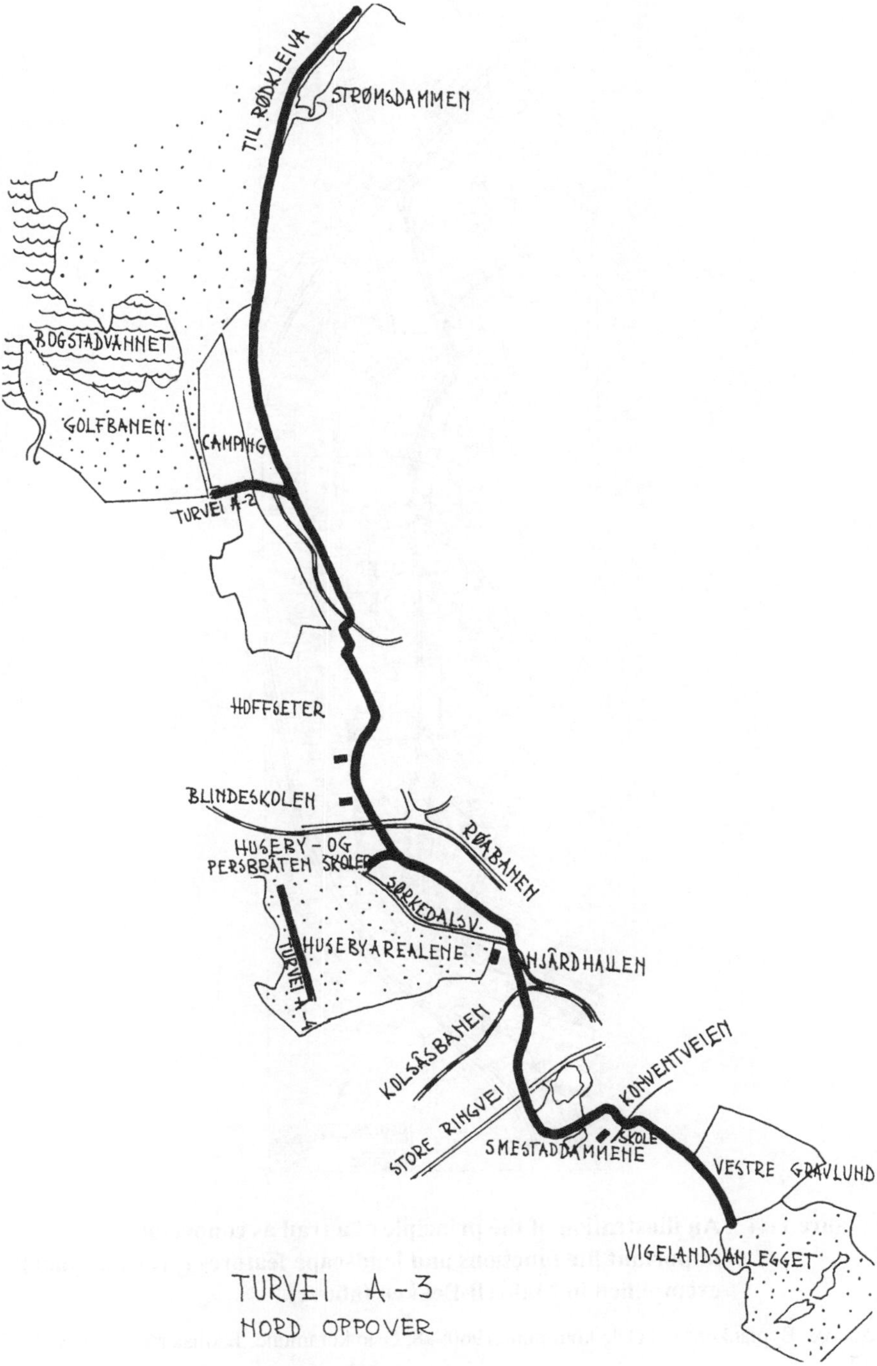
TIL RØDKLEIVA
STRØMSDAMMEN
BOGSTADVANNET
GOLFBANEN
CAMPING
TURVEI A-2
HOFFSETER
BLINDESKOLEN
HUSEBY - OG PERSBRÅTEN SKOLER
RØABANEN
SØRKEDALSV.
HUSEBYAREALENE
TURVEI A-1
NSÅRDHALLEN
KOLSÅSBANEN
KONVENTVEIEN
STORE RINGVEI
SMESTADDAMMENE
SKOLE
VESTRE GRAVLUND
VIGELANDSANLEGGET
TURVEI A-3
NORD OPPOVER

Figure 11.4 An illustration of the principle of a trail as connector of important life functions and landscape features (previous page); exemplified in Makrell-Bekken (above)

Source: Byplankontoret Oslo kommune 1960: 48; Oslo kommune, Teknisk rådmann 1967: 17.

an extra boost to that end, and also contributed to securing land areas along the coastline. There are clear indications in all three of the municipal master plans for Oslo between 1950 and 1980 that this important planning work linked the city's green assets together.

The green plan principles on which planning during the period between 1948 and 1980 are based can be said to reflect influences from the USA, England and Sweden. The trail plan was a continuation of Harald Hals's parks system concept, which in turn has clear connections to Olmsted's line of thought. According to Kvarv (2003), the ideas of sub-centres and neighbourhoods linked by green corridors came from England and Sweden, and the principles were already presented in the municipal master plan from 1950. With this plan, according to Ellefsen (1992), functionalism's urban plan ideals had also gained support in Norway, but to Kvarv's mind (2003) the garden city element was also essential. Good public transport service was also part of the concept. Throughout the whole period between 1948 and 1980 strong political and social engagement lay behind the idea of the green spaces, with an emphasis on health, well-being, recreation and outdoor life. The plan from 1980 expressed this in the following way: 'The need for outdoor recreation is important for reasons of health and well-being and must be given the same weight as housing, employment and services' (Oslo kommune 1980: 41). The distribution aspect was also given weight, as was the protection of cultivated and cultivable land, and towards the end of the period green values were put in a natural resource and environmental perspective (Oslo kommune 1980).

1980–2009: Discontinuation and Decline

The problems that had accumulated around the end of the period mentioned above did go away. The decline in the population growth rate continued, which in turn adversely affected tax revenues and the municipality's finances. A general economic recession combined with new general conditions from the government did not help the situation. Added to that, there was a decline in some parts of the housing market, due partly to high prices but also because the municipal master plan from 1984 (Oslo kommune 1984) had not aligned housing development with market demands. To improve the municipality's finances and afford it some leeway, the municipal plan in 1984 emphasized concentrating efforts on increased housing development and population growth. The latter of these in particular required a shift away from the policy of the preceding period when the aim had been to reduce the rate of population growth in Oslo.

The strategy for land use policy that was determined in 1991 is the one which Oslo continues to pursue: densification inside existing town boundaries (Byrådet Oslo kommune 1991). The trend of urban expansion and land use expansion was therefore discontinued. The decline in the population growth rate abated in the mid-1980s, and since 1986 Oslo has grown, in some periods at a faster rate than the municipality's own prognoses indicated. The provision of sufficient housing has therefore been an important issue in the municipal plans for the 2000s.

Since the 1990s discussions have to a large extent been about where densification should occur and at what rate. In the municipal master plan for 2000 (Oslo kommune 1999), two different densification strategies were examined: housing developments along the fjord, and housing developments along the rail network, and the plan proposed a combination of these two alternatives. This strategy has been adopted in planning work right up to the present day, and housing development has to a large extent occurred in connection with transport hubs along the rail network though with an emphasis on the inner city area. (Oslo kommune 2004; Oslo kommune 2008). In other words, the most densely developed areas in Norway have become even more so, and the 'depopulation' policy aimed at reducing welfare problems caused by high levels of densification from the 1960s and 1970s is definitively a thing of the past.

In addition, during the 1980s a series of political reforms were introduced which have had various impacts on the city's green spaces. These include a trio of measures introduced in 1991:

- The reform of the urban districts, with power decentralized to the local level. Green spaces were mentioned as a suitable example for full decentralization but were not followed up with concrete measures (Oslo kommune Byrådet 1991).
- Competitive tendering of municipal services, including operation of the green spaces. The reform was commenced in the early 2000s when the Parks Department's services were first put out to tender and were subsequently fully privatized in 2011.
- Reorganization of the municipality's property policy. Where it previously advocated the extensive acquisition of property during the post-war period in order to facilitate urban development, including securing trails and green spaces, the municipal master plan of 1991 now cleared the way for extensive disposal of property (Oslo kommune Byrådet 1991).

The character of the overall planning also changed, and the municipal master plan of 1991 was presented as a strategic steering document (Oslo kommune Byrådet 1991). The land use plan of 1991 was therefore no longer a comprehensive and legally binding plan for all of the city's land areas, but was instead a set of principles for the way in which it was desired that the city as a whole should be developed. Separate municipal sector plans which dealt with specific subject areas, such as green structures, or with specific geographical areas of the city were to be used to manage the details of land use development.

The following expression of frustration, which appeared in the book issued to commemorate the 75th anniversary of the Parks Department, suggests that the work on Oslo's parks and trails was at low ebb:

> Everyone who is supposed to safeguard the functions of the trails and see the holistic aspect of the system is often frustrated over all the 'good causes' such as housing for the elderly, day-care centres, expansion of car parking facilities, etc. that require smaller or larger sections of the trails. Areas that are regulated for public use, such as trails, parks and large outdoor recreation areas, are inadequately protected and are vulnerable to rezoning. (Brenøe 1991)

Even in the supporting documentation for the green plan that was in the pipeline, the city planning office documented that the parks and trail system was no longer prioritized and had begun a process of fragmentation (Oslo byplankontor 1990).

Since the master land use plan became merely a strategy document from 1991, details were managed through more binding municipal sector plans. In principle, the general 'public' green structure that was dealt with in the Municipal Plan for Green Structure in the Building Zone was approved in 1993 (Oslo 1991). The private green structure was to be governed by the Zoning Plan for Small Houses in the Outer City of Oslo of 1997 (Oslo bystyre 1997). In addition, the overall green strategies are today decided by the City Council in the Urban Ecology Programme that was approved for the first time for the period 2002–14 (City of Oslo City Government 2003).

Forty years would pass, from 1953 to 1993, before a new general plan for Oslo's parks and green spaces was created. As with the trail plan of 1953, the objective for the new plan was to join together parks, green spaces and, not least, the green belt, to the extensive trail network. The green plan is primarily referred to as a general plan that sets out the principles for the green structure in general. Under the Planning and Building Act, land use is only legally binding for the areas with greatest strategic significance, whereas the plan sets out guidelines for sub-areas and types of land use, for example residential areas and green spaces along the transport arteries.

In addition to recreational activities, this plan should also take into account aesthetical aspects of the landscape and should also safeguard ecological assets. The latter of these was a new issue for green planning in the 1990s. The report on ecological land use principles for the building zone in Oslo was therefore ground-breaking in terms of safeguarding biodiversity in urban areas in Norway (Nyhuus 1991). With this plan, Oslo led the way in introducing the urban green structure as a multifunctional element in an urban environment, which resulted in the City of Oslo being awarded the Norwegian Housing and Urban Planning Association's Urban Planning Award for 1993.

The green plan of 1993 restored greenness as a main planning topic and contributed to new knowledge about the values and functions of green structures for more than just recreational purposes (Oslo kommune Plan- og bygningsetaten 2009). The plan was subsequently followed up with more detailed planning for sub-areas, for example Groruddalen (Oslo kommune Friluftsetaten and Viak 2003).

In addition to the green plans, the municipality undertook extensive planning work during this period on the Akerselva Environmental Park (Oslo byplankontor 1989), a project which, we have established, was begun in the early 1990s. In the

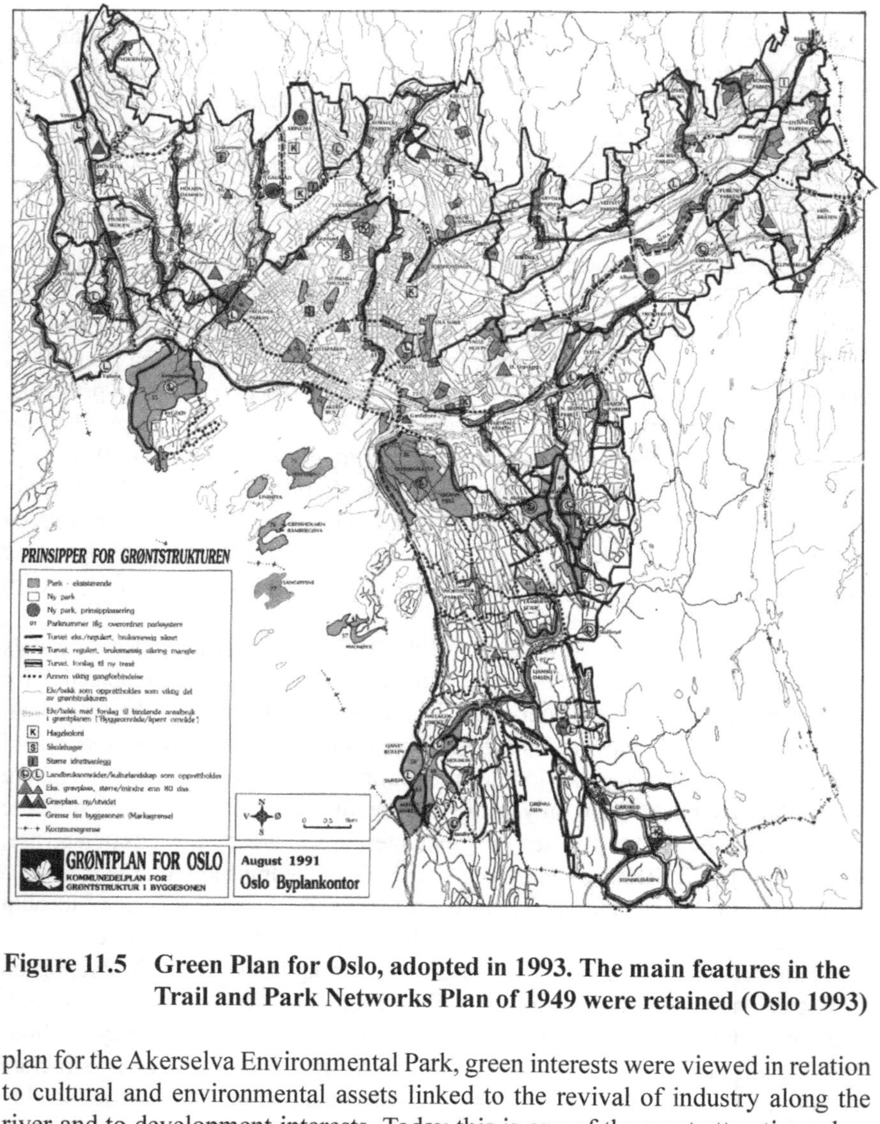

Figure 11.5 Green Plan for Oslo, adopted in 1993. The main features in the Trail and Park Networks Plan of 1949 were retained (Oslo 1993)

plan for the Akerselva Environmental Park, green interests were viewed in relation to cultural and environmental assets linked to the revival of industry along the river and to development interests. Today this is one of the most attractive urban recreational areas in Oslo. Similar work is underway in Alna, one of the city's other important stretches of river (Oslo kommune Plan- og bygningsetaten 2009).

As we have seen, there was no lack of overall planning for green purposes between 1980 and 2009. Nonetheless, this has not prevented green areas from being developed. A study on the impact of densification on Oslo's green structure revealed that large contiguous vegetation-covered areas had been reduced in size or fragmented (Guttu 1997). Moreover, the content had been more cultivated, and the original vegetation had had to make way for more garden-like vegetation.

Tree-covered areas in the building zone had also been reduced in size. More recent studies from Oslo show the same trend (Thorén, Due Trier et al. 2010).

The municipal master plan of 2008 revealed the same trend for public green spaces that were reduced from 49 square metres per inhabitant in 1998 to 47 square metres in 2006, despite more public green spaces in total (Oslo kommune 2008). The reason for this is the significant growth in the population. There are also significant differences between the inner and outer city areas: inhabitants in the inner city area had access to approximately half the size of green space (30 square metres) compared to that of the inhabitants in the outer city areas (55 square metres). Despite the fact that the municipal master plan of 2008 revealed a lack of areas for sports and physical activities and of natural and green spaces in the inner city, it was proposed that most of the development should take place there, with no discussion of this dilemma in the planning document.

The development of green spaces, both regulated and otherwise, and the lack of any long-term strategy for conserving the large green spaces, has given rise to frustration among a number of Oslo's inhabitants. In the course of the period we have covered here, several green, grassroots movements were formed in Oslo, such as the Network for the Protection of Green Lungs and the Oslo River Forum, with campaigns to open up stretches of rivers and streams that have been closed.

The period between 1980 and 2009 is full of paradoxes when it comes to Oslo's green spaces. On the one hand, we have shown that green and vegetation-covered areas were reduced and that there are fewer square metres of publicly accessible green spaces per person. On the other hand, there has hardly ever been greener, municipal planning activity in the city's history – without any apparent benefit to the population in the form of more or larger green spaces. One important reason is that the municipality's change of policy from depopulation and urban-rural migration to one of increased population and densification of the city. The main arguments for the change of policy were, as mentioned earlier, to boost municipal finances and gain some financial leeway.

This policy pre-empted the government guidelines for urban development in Norway based on densification, which came a little later in the Report to the Storting no. 31 (Miljøverndepartementet 1993). The government's arguments for justifying densification took a broader perspective. The goal that was presented in the Report to the Storting no. 31 was to shift land use policy in a more sustainable direction by, for example, following up Norway's obligations in the wake of the UN environmental conference in Rio de Janeiro in 1992 (Miljøverndepartementet 1993). Densification should contribute to reducing private car use and, consequently, CO_2 emissions, and prevent the loss of biodiversity. The parliamentary report also stressed the necessity to safeguard areas of land that were important for food production. For Oslo's part, the sustainability arguments justifying densification were not used until much later, for example in the municipal master plan of 2000 (Oslo kommune 1999). Today, climate change arguments and sustainable development have become two of the main arguments for justifying the densification policy (Oslo kommune 2008).

Norwegian densification policy was also influenced by the research project Nature and Environmentally Friendly Urban Development (Næss 1992) which was going on during the 1980s and which was based on a sustainable development in line with the Brundtland Report (World Commission on Environment and Development 1987). The urban green structure was incorporated as an essential element in the new policy because green spaces could contribute to both unfavourable and extensive urban growth, while at the same time such spaces are important to the health and well-being of future inhabitants of more densely populated cities. Green structure planning was therefore regarded as vital in urban development, something which is also evident in the Report to the Storting no. 31 (Miljøverndepartementet 1993).

In other words, the revival of green planning in Oslo in the 1990s can be viewed in the light of government policy and of the above mentioned research project. The municipality itself had a tradition dating back to the 1920s for planning green spaces using a systematic approach. What was new in the green plan of 1993 was the multifunctional approach. This originated in the focus at that time on sustainable development, where biodiversity, for example, was a central element, something to which the abovementioned NAMIT project also attached weight (Næss 1992). A multifunctional green structure implies that green spaces have more assets/functions than the traditional green spaces for recreational purposes; for example biological, local climate, for the hydrological system and so on. Moreover, multifunctional green structure is a topic which enjoys widespread international support today (Werquin et al. 2005).

The municipality's focus on economic and market aspects has also influenced the arguments for greenness. Oslo's competitive advantage both nationally and internationally has become more important. The municipality therefore regularly participates in international competitions, and in 2003 it was awarded the European Sustainable City Award. This award is actively used in selling the city abroad. For example, the following is an extract taken from the municipal master plan of 2008: 'Oslo also has a distinctive advantage as an attractive environmentally friendly city surrounded by the fjord and the forest, and in 2003, in fierce competition with 60 other cities, it was voted Europe's sustainable city' (Oslo kommune 2008: 5). High scores for good accessibility to natural and recreational areas was an important reason. Today 'the blue and the green and the city in between' is an important trademark for Oslo.[5]

Today's Green Structure

Since the new Planning and Building Act came into force in 2009, the City of Oslo had to change its practice for handling strategic plans. The new act required a legally

5 The master municipal plan of 2000 introduced the term 'blue-green profile' Oslo kommune (1999). *Kommuneplan 2000: arealdel.* Oslo, Plan- og bygningsetaten.

binding land use plan for the entire municipality. The idea behind a collective land use plan was, among other things, to create more predictability and, not least, to obtain a better overview for the municipality's inhabitants. This is an important change from a green perspective. The strategic plans mentioned the green structure in a non-committal manner, and one had to plough through numerous municipal sector plans, thematic and geographical, to obtain a full overview. The legally binding document from 2009 is in reality only a map summary of the most important binding plans that were approved, and the green plan from 1993 is not mentioned. It is only in the text that we are informed that this plan also applies.

Moreover, one could argue that the green plan from 1993 is out of date. One criticism of the plan is that it is too non-committal and loosely formulated. A new draft plan is therefore undergoing consultation (Oslo kommune Plan- og

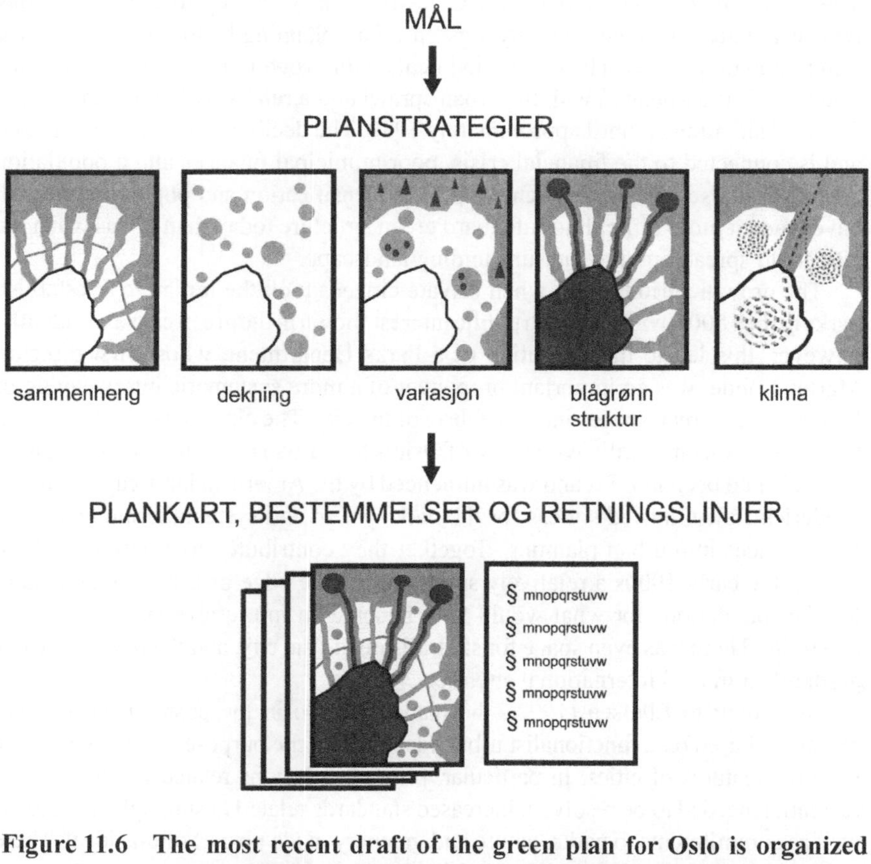

Figure 11.6 The most recent draft of the green plan for Oslo is organized around five main themes: coherence, coverage of green areas, variety, blue-green structure and climate

Source: Oslo kommune, Plan- og bygningsetaten 2009: 11.

bygningsetaten 2009). This draft is legally binding to a greater extent, and several small green spaces in the city districts are included.

The green structure as a multifunctional system in the city is further consolidated in the new draft of the green plan. Biodiversity as well as climate changes and the importance of green structure in that context are given more attention. The blue-green structure is also an important topic that follows up on Oslo's slogan, 'The blue and the green and the city in between', and international directives such as the EU's water directive. Figure 11.6 is a conceptual representation of how the new green plan is structured.

Summary and Conclusion

The review of the master plans for Oslo from the 1900s up to the present day reveals a pattern whereby the city's green urban planning history can be divided into two main periods. The one period deals with growth, the other with decline. The growth is associated with the urban sprawl and a relatively healthy municipal financial situation up until approximately 1980. The decline comes at around 1980 and is connected to the financial crisis, poor municipal finances and a population decline, but also to a shift in general policy. Densification and population growth have made it more difficult to safeguard green structure today than it was when the city could spread out into the surrounding landscape.

The growth period began when private citizens took the initiative to establish parks in the 1800s, without much public interest shown in the green cause. Gradually, however, this led to the formation of a Parks Department, whose first director, Marius Røhne, was an important proponent of a more systematic interweaving of both blue and green spaces into the fabric of the city. The circumstances were such that his ideas seem to tally with those of Oslo's first urban planning director, Harald Hals. He had been in USA and was influenced by the American landscape architect Frederick Law Olmsted's ideas for the park system and systematic integration of green spaces into urban planning. Together they contributed to a narrative about how in the early 1900s a relatively small city on the edge of Europe managed to lay the foundations for what would later become an impressive system of parks and trails. There was even space for ski trails inside the city, and the planning work gradually attracted international attention.

According to Ellefsen (1992), the plans for Oslo in the post-war years were originally based on a functionalist urban ideal, where the purpose was to change the negative features of cities. In particular, primary functions related to housing and recreation needed to be resolved. Increased standards related to sun, light, air and, as well as commitments to parks, nature and greenery made up part of this ideal. There was a desire to behold the urban-rural boundary and protect the green belt around Oslo from development. Harald Hals's parks system was continued in the trails plan approved in 1953. This planning was strengthened and rendered visible in the master plans right up until the mid-1970s. Raising standards in the central residential areas were also given priority. All this was possible thanks to the expansive public land acquisition, public initiatives in housing policy and, not least, comprehensive municipal planning, which included green planning objectives. Nonetheless, the fundamental precondition had been the land use expansion that took place when the Oslo and Aker municipalities were merged in 1948.

The 1980s represented a vacuum for green planning, as well as a reorientation in the view of how the municipality should deal with growth and population growth. It was a period of economic decline, and resources for green spaces, such as safeguarding the trails, were scarce. The policy was therefore changed in the early 1990s, and the objective became population growth combined with densification. It is a paradox that this entailed a shift in the direction towards green planning, which once again came into favour after 40 years.

This period was full of paradoxes. Along with the increased focus on green planning and intensive production of plans for green purposes, Oslo's green structure became fragmented and at times lost to development. The growth in the population meant that there were fewer square metres of regulated green space per person, despite the fact that the municipality had allocated more land. The inhabitants with the smallest green space per person were those who lived in the central districts of the city, precisely where today's politicians want to see further densification. They claim that there is sufficient space for urban growth, yet at the same time complain about the lack of spaces for play, recreation and sports. During this period there has been large-scale selling-off of municipal land, in spite of the scarcity that was highlighted.

Combined with the population growth, the densification policy is an important cause of the current situation, with development of green assets and grassroots movements that are fighting for their local landscape and woodlands. The fragmentation of the green structure can also be viewed in the light of the City of Oslo's fragmented planning system. Management through strategic plans in combination with a mountain of sub-plans has made it difficult for the inhabitants of the city – and perhaps also for the politicians – to gain an adequate overview. Moreover, a lot is left up to the market and to negotiations with developers. This makes it even more difficult for the general public to keep tabs on what goes on. The amendments in the new Planning and Building Act that was enforced in 2009 are intended to rectify this situation when the municipal master plan comes up for revision in a few years' time.

The Arguments

Throughout the entire period covered by this study, the main arguments for green spaces in the city have been health and quality of life. Another argument which gradually came into play was the edifying role and the educational importance of greenness. Aesthetics and urban landscape were early arguments for greenness in the city. According to the master plan of 1950, green spaces should both unify the city and divide it into smaller units. Furthermore, the trails should be adapted to the terrain, and areas of particular significance to the landscape should be preserved. The small house plan is one example of how the green character must be preserved, even if the residential belt is densified. Another point is that today great importance is attached to the multifunctional significance of the green structure, and many assets and functions have emerged since the idea of the parks system was introduced in the 1920s. Greenness as a multifunctional system is therefore a new concept and rhetorical tool that may prove to be a powerful one in the further development of Oslo's green structure.

The plans we have reviewed show that international agreements, treaties and celebratory years of various kinds have had as much to say for the argumentation, as have the realities of the plans. A consequence of the introduction of nature conservation legislation in the 1970s was that the plans directed attention to maintaining biodiversity. The UN's annual Children's Day helped to sharpen the focus on children's use of the city. The climate change issue of recent years has, first of all, contributed to the densification policy but also to arguments for the need for more greenness inside the city. Another argument that has emerged along with market thinking is the of selling Oslo with the help of slogans. Not least, this is part of a strategy to sell Oslo, both nationally and internationally.

Nonetheless, Oslo has a history of visionary urban planning in the first half of the 20th century to thank for the structure of the blue and green spaces and trails which the city enjoys today. This goes to show how important it was that someone thought systematically and conceptually about how such a system of parks and trails should look. There is still strong engagement in green issues in Oslo. Also, the scarcity of and competition for land has gradually given rise to public pressure in defence of green spaces, something which the Oslo River Forum and the Network for Green Spaces illustrate.

Bibliography

Apall-Olsen, I.A. 2007. *Norsk hagekunst under funksjonalismen: Idelologisk bakgrunn og estetisk praksis i hager og parker*. Oslo: I.A. Apall-Olsen: 104 bl., ill.

Aspen, J. 2010. Urban planning as representation: An examination of Harald Hals' general plan for Oslo: S. 271–89, in C.H. Cordua, *Manifestoes and Transformations in the Early Modernist City*. Farnham: Ashgate.

Berg, A. 1965. *Det gamle Christiania*. Oslo: Dreyers forlag.

Brenøe, B.-A. 1991. Turveier – grønne korridorer. *Den grønne by*. I. o. T. Tønnessen, Dagfinn (ed.). Oslo, Grøndahl & Søn Forlag A.s, Oslo kommune Park- og idrettsvesenet, Det norske hageselskap: 83 –7.

Bruun, M. 1984. Streiftog i Oslos gamle hager og parker. *St. Hallvard* 3(3/84): 8.

Bruun, M. 1992. Parker og grønnområder i byen. *Årbok: (Foreningen til norske fortidsminnesmerkers bevaring)*. 146.

Byplankontoret Oslo kommune. 1960. *Oslo: Planlegging og utvikling: Oversikt over den geografiske og historiske bakgrunn, utviklingen av befolkning og næringsliv m.m. og planlegging og utbygging etter krigen*. Oslo: Oslo kommune.

Byplankontoret Oslo kommune. 1960. *Oslo: Planlegging og utvikling: Oversikt over den geografiske og historiske bakgrunn, utviklingen av befolkning og næringsliv m.m. og planlegging og utbygging etter krigen / utarbeidet ved Byplankontoret*. Oslo: Oslo kommune.

Byrådet Oslo kommune. 1991. Kommuneplan 1991 Strategi for økonomisk handlefrihet.

City of Oslo. 2003. *Strategy for Sustainable Development Environment and Sustainability Status 2002 Urban Ecology Programme 2002–14*. Oslo.

Ellefsen, K.O. 1992. *Årbok (Foreningen til norske fortidsminnesmerkers bevaring)* 146: s. 81–95: ill.

Fabos, J.G. 1995. Introduction and Overview – the Greenway Movement, Uses and Potentials of Greenways. *Landscape and Urban Planning* 33(1–3): 1–13.

Fabos, J.G. and Ryan, R.L. 2006. An introduction to greenway planning around the world. *Landscape and Urban Planning* 76 (1–4): 1–6.

Generalplankomitéen for Stor-Oslo 1934. *Stor-Oslo: Forslag til generalplan fra den av Oslo, Aker og Bærum kommuner under 24-11-1928, 23-10-1929 og 17-3-1932 nedsatte komité: Karter og bilag*. Oslo: Komitéen.

Guttu, J. 1997. *Boligfortetting i Oslo: Konsekvenser for grønnstruktur, bokvaliteter og arkitektur*. Oslo: Norsk institutt for by- og regionforskning.

Hals, H. 1929. *Fra Christiania til Stor-Oslo: Et forslag til generalplan for Oslo*. Oslo: Aschehoug.

Hansen, T. and Guttu, J. 2000. *Oslo kommunes boligpolitikk 1960–1989: Fra storskalabygging til frislepp*. Oslo: Norsk institutt for by- og regionforskning Norges byggforskningsinstitutt.

Hirschfeld, C.C.L. (ed.). 1779–85. *Theorie der Gartenkunst*. Leipzig: M.G. Weidmanns Erben und Reich.

Howard, E. 1902. *Garden Cities of To-morrow*. London, Swan Sonnenschein & Co.

Jensen, R.H. 1980. *Moderne norsk byplanlegging blir til: Tanker og ideer som preget fremveksten av moderne norsk byplanlegging slik det særlig fremkommer i de tekniske tidsskriftene 1854-1919 med forenklet videreføring av vesentlige tendenser i 1920–30 årene*. Trondheim: [s.n.].

Jørgensen, K. and Fredriksen, B.A. 2008. Folkepark og kongsgård. Bevaring og utvikling på Bygdøy. *Arkitektur, N* 06/2008.

Kvarv, S. 2003. Yrkesroller og fagideologiske brytninger i fysisk planlegging i Norge, 1920–70. Oslo: Arkitekthøgskolen i Oslo, 12: 241 s.

Lexau, S.S. 1993. Harald Hals' generalplanforslag fra 1929 'Fra Christiania til Stor-Oslo': Sentrale estetiske aspekter ved norsk byplanlegging 1920–1940: En kunsthistorisk analyse. Bergen: S. S. Lexau.,140 bl., ill.

Miljøverndepartementet 1993. Den regionale planleggingen og arealpolitikken (Regional planning and spatial policy). Miljøverndepartementet. Oslo: Miljøverndepartementet (Ministry of Environment).

Moland, T. 2006. *Historien om Nordmarka: Gjennom de siste 200 år*. Oslo, Christiania forlag.

Nyhuus, S. 1991. *Økologiske arealprinsipper i Oslos byggesone*. Oslo: Oslo, Byplankontoret.

Næss, P. 1992. Natur- og miljøvennlig tettstedsutvikling. Faglig sluttrapport (Environmentally Sound Urban Devellopment. Final report). *NIBR reports*. NIBR. Oslo: Norwegian Institute for Urban and Regional Research, NIBR report 1992: 2.

Næss, P. 1992. Natur- og miljøvennlig tettstedsutvikling: Faglig sluttrapport (Environmentally Sound Urban Development. Final Report. *NIBR – rapport*. N. i. f. b.-o. regionforskning. Oslo: Norsk institutt for by- og regionforskning, 407.

Oslo kommune Teknisk rådmann 1967. Turveier i Oslo (Trails in Oslo). Oslo: Emil Moestue A.S.

Oslo. 1975. *Oslo kommuneplan 1976–1985: Fylkes- og generalplan: Administrasjonens forslag*. Oslo: Kommunen.

Oslo. 1991. *Grøntplan for Oslo: Kommunedelplan for grøntstruktur i byggesonen: høringsutkast september 1991*. Oslo: Oslo byplankontor.

Oslo byplankontor. 1989. *Akerselva miljøpark: Forslag til kommunedelplan*. Oslo: Byplankontoret.

Oslo byplankontor. 1990. Grønnstruktur i byggesonen. *Feltnotat*.

Oslo bystyre (Oslo City Council). 1997. Reguleringsplan for småhusområder i Oslo ytre by (Småhusplanen) (Zoning plan for areas outside Oslo city). Oslo.

Oslo kommune. 1980. *Oslo kommuneplan 1980–1990*. Oslo: Oslo kommune.

Oslo kommune. 1984. *Oslo kommuneplan 1984*. Oslo: Oslo kommune.

Oslo kommune. 1999. *Kommuneplan 2000: Arealdel*. Oslo, Plan- og bygningsetaten.

Oslo kommune. 2004. *Kommuneplan 2004: Oslo mot 2020: Vedtatt av Oslo bystyre 26.05.2004 (sak 255)*. Oslo: Oslo kommune.

Oslo kommune. 2008. *Kommuneplan 2008: Oslo mot 2025: Vedtatt av Oslo bystyre 11.06.2008 (sak 213)*. Oslo: Oslo kommune.

Oslo kommune Byrådet. 1991. *Oslo kommuneplan 1991, Strategi for økonomisk handlefrihet Arealplan*. Oslo.

Oslo kommune Friluftsetaten and A. Viak. 2003. *Strategisk grønnstrukturplan for Groruddalen. Foreløpig – Mars 2003*. Oslo.

Oslo kommune Plan- og bygningsetaten. 2009. Grøntplan for Oslo Kommunedelplan for den blågrønne strukturen i Oslos byggesone. P. o. bygningsetaten. Oslo: 76 s + vedlegg.

Oslo kommune Plan- og bygningsetaten. 2009. *Kommunedelplan for Alna miljøpark*.

Rolfsen, E. 1950. *Generalplan for Oslo: Et utkast lagt fram som diskusjonsgrunnlag for de kommunale myndigheter og etater og for andre interesserte*. Oslo: Oslo reguleringsvesen.

Røhne, M. 1967. *Oslo kommunale parker og grønnanlegg 1810–1948*. Oslo. D.2. De enkelte anlegg.

Statistics Norway. 2011. Population statistics. Population changes in municipalities. 1951–2011. 03 03 Oslo. Population 1 January and population changes during the year. 1951–. Available at: http://www.ssb.no/emner/02/02/folkendrhist/tabeller/tab/03.html (accessed: 20 September 2011)

Thorén, A.-K.H. and Nyhuus, S. 1994. Planlegging av grønnstruktur i byer og tettsteder (Planning for urban green structures). *DN-Håndbok*. D. f. naturforvaltning. Trondheim: Direktoratet for naturforvaltning. 6.

Thorén, K.H. et al. 2010. Kartlegging av urban grønnstruktur med satellittdata (Development of a service for monitoring the urban green land cover). *Kart og Plan* 4.

Tvinnereim, H.S. 1990. *Norsk byplanlegging 1920–1965: Nasjonal tradisjon, internasjonal påverknad*. Volda: Senter for kommunalforskning.

Tønnesen, I.F. and Tveito, D. 1991. *Den Grønne by: Oslo Park- og idrettsvesen gjennom 75 år: 1916–1991*. Oslo: Grøndahl.

Utviklings- og kompetanseetaten and Satistics Norway 2009. Population 1801–2009 Christiania/Oslo and Aker. *Statistisk aarbok Oslo*. Available at: http://www.utviklings-og-kompetanseetaten.oslo.kommune.no/getfile.php/utviklings-%20og%20kompetanseetaten%20%28UKE%29/Internett%20%28UKE%29/Dokumenter/Oslostatistikken/ARKIV/statistikk/Statistisk%20%C3%A5rbok%202009/Statistisk%20aarbok%20for%20Oslo%202009%20kap%2002.pdf (accessed: 20 September 2011).

Werquin, A.C. et al. 2005. *Green Structure and Urban Planning: Final Report*. Luxembourg: Office for Official Publications of the European Communities.

World Commission on Environment and Development. 1987. *Our Common Future*. Oxford: Oxford University Press.

Chapter 12
Planning a Compact Oslo

Eva Falleth and Inger-Lise Saglie

Introduction

The compact city has been put forward as a way of obtaining sustainable cities in Norway for almost 20 years. This ideal has been implemented in national as well as local policy. The core ideas are to increase the density within already built-up areas in order to reduce the need for cars and therefore the emission of greenhouse gases, and to protect green areas in urban environments (Ministry of the Environment 1992, 1996, 2001). Finding a balance between these contradictory goals is a key requirement for developing successful compact cities.

The compact city policy has coincided with a neo-liberal movement in planning. Planning is no longer an exclusively public activity; most local development plans are now devised by private actors.[1] In Norway, unlike most countries, private actors are entitled to submit draft development plans[2] for political approval by the local government (Falleth et al. 2010, Kalbro et al. 2010). A total of 68 of the 88 development plans which were adopted in Oslo in 2009 (Norway Statistics 2011) were submitted for political approval by private actors. An application for a building permit is often closely integrated with these plans such that permits have been labelled as 'project plans'. Project plans are specifically tailored to a particular building project. They form a patchwork pattern of plans, due to their miniature format on small plots within urban built-up areas (Børrud 2005, Røsnes 2005, Nordahl 2006). Bottom-up private project plans have challenged the traditional logic of the hierarchical planning system. Project plans appear to be more dependent on market mechanisms than master planning.

In practice, project plans have often been in line with the compact city ideal of high densification (Falleth and Saglie 2011). However, the rationale for densification in project plans and the compact city ideal are different. Project plans favour high density as it results in higher profit, while the compact city ideal favours high density as a result of environmental concerns. Despite different rationales for densification, the final output is often the same: high-density development projects. This mix of economic and environmental arguments can easily favour specific

1 Private plans might be submitted by private and public actors. Norway Statistics refer to both kinds as private plans. Most of them are submitted by private companies (Børrud 2005, Røsnes 2005).

2 Development plans are formal zoning plans (*reguleringsplan*).

environmental concerns, while ones that are more difficult are left unaddressed. We believe that the protection of urban green areas, which makes development on such sites impossible, is one such aspect that may be ignored.

The combination of environmental arguments and economic interests among market actors and professionals provides a convenient way for politicians to argue for development as well. Politicians are dealing with a shortage of urban dwellings in a rapidly growing city. Today, in 2011, Oslo has approximately 600,000 inhabitants, and is one of the fastest growing cities in Norway. It is predicted that there will be around 800,000 inhabitants in Oslo in 2030 (Norway Statistics 2010). This population growth has led to a great deal of pressure on housing stocks and demands for development. Politicians and developers can therefore easily agree on plans for urban development, while the role of advocate for the protection of green areas is left to other actors.

The priority given to densification may result in a bias towards climate-related concerns. The reduction of greenhouse gases which results from high density and a reduced need for transportation fits well with current urban development practices and developers' arguments for high-density areas. The protection of green areas may well receive less attention, specifically among developers. There is a lack of knowledge of the consequences of densification for green infrastructure, but current research on the loss of urban green infrastructures is alarming (Thoren 2010). This chapter describes the arguments for and against urban development. We will investigate the extent to which the protection of urban green infrastructure and densification are regarded as important in urban development. We also intend to study the strength of arguments in practice which have been developed in order to protect green urban infrastructure. Finally, we will try to explain why some development projects in green areas are stopped while others are not.

We will address these issues in two ways. First, we will analyse the role of green infrastructure within the Norwegian professional discourse on the compact city model. This will be analysed by a study of the contributions on urban sustainability in the journal *PLAN*. *PLAN* is the main journal for the dissemination and discussion of information on planning in Norway, and is widely read among academics as well as practitioners. The professional debate in *PLAN* influences both the norms and the rules which are related to political-administrative control, as well as the content of private plans provided by professional planners. Second, we will analyse the use of arguments in planning practice. We will analyse the arguments for and against development in two projects in Oslo, which conflicted with maintaining the integrity of the green infrastructure: the Geitmyra School Garden and Tiedemann Playground cases. The resolution of the cases led to different outcomes. The Tiedemann Playground development was realized, while work on the Geitmyra School Garden was stopped. These cases have not been chosen to be representative of all development proposals, but to illustrate the range of arguments for and against development. Therefore, the findings cannot be generalized. These two cases were chosen from a list of 22 development projects, which threatened the Oslo area's established green infrastructure (Groenne lunger 2010). Two major kinds of development project dominate this list: private project plans for housing development and public plans for the development of kindergartens. We have chosen one of each kind. The kindergarten in question was initially planned to be a residential area.

We will also build on a research project on democracy in neo-liberal urban planning. This research included a broad survey of planners, politicians, developers and civil society organizations, a document study of 100 development plans (Falleth et al. 2008), case studies of development projects (Netland 2008, Windju 2008) and qualitative interviews among planners, politicians, developers and civil society organizations (Hanssen and Saglie 2010, Hanssen 2010).

The Compact City Discourse

The compact city ideal can be traced back to the report of the World Commission on Environment and Development publication of 'Our Common Future' in 1987 (World Commission on Environment and Development 1987). In Norway, the important research project 'Nature and environmental friendly urban development' (NAMIT) was set up, inspired by this report. The aim of NAMIT was to develop criteria for sustainable urban development, and through the use of different scenarios, to assess which patterns of urban development would be sustainable (Næss 1992). The core scenario involved a concentrated pattern of development, a restructured transport system geared towards public transport and the protection of green areas. The urban ideal was densification in the city as a counteraction to pre-existing urban sprawl. The NAMIT scenario was later renamed the 'compact city' ideal, and was in line with international trends (Breheny 1992, Jencks et al. 1996). NAMIT has influenced urban planning ideals and planning policy to the extent that it can be said to be hegemonic in Norwegian planning discourse (Næss et al. 2009).

We have investigated which elements within the compact city scenario have received attention from professionals in *PLAN*. *PLAN* is widely used as a source for the analysis of professional discourse when writing the history of Norwegian planning (for example Næss et al. 2009, Kleven 2011). A total of 1,000 contributions appeared in *PLAN* in the period 1990–2010. Around 45 of them referred to urban sustainability, and 15 of these can be related directly to the compact city ideal. Only three of the 45 contributions addressed the issue of green infrastructure, while 13 contributions addressed sustainable mobility within the climate discourse. The 'short cycle' city, which emphasizes short flows of, for instance, energy, water and local food production, was dealt with in only two of these articles, both published in the 1990s. This 'short cycle' perspective has been totally absent in the 2000s. The remaining articles about sustainability have either a general focus on sustainability, describe concrete development projects or address the rising consumption levels. These figures indicate that professionals have a clear tendency to give priority to mobility and climate-related challenges, rather than to the green infrastructure. In addition, national policy has followed the ideal of a compact city, including national policy guidelines giving priority to densification and integrated land use and transport planning.

Mobility, densification and climate-related issues are, however, of great importance. National policy guidelines give priority to integrated land use and transport planning, compact cities and densification. The 'White Paper on Sustainable Development' emphasizes public transport as the backbone of sustainable urban development (Ministry of the Environment 1996). The White Paper also focuses on the development of urban centres with a concentration of workplaces, dwellings, shops and cultural facilities close to public transport nodes. It also highlights the protection of the green infrastructure and large natural areas. While transport and green areas provided the initial justification for compact cities, there is today a growing focus on compact city development from architectural/urbanism perspectives (Næss et al. 2009). None of these aspects are new to urban planning, but they have all merged into the idea of the compact city, making this ideal hegemonic, comprehensive and unassailable.

The national guidelines as promoted by the Ministry of Environment stress the importance of high density and the green infrastructure. High density is, in a way, a prerequisite for the preservation of green areas, in much the same way as Le Corbusier wrote in *The City of Tomorrow and its Planning* (1925/1971). However, it was acknowledged that arguments for densification, in practice, could be a threat unless measures are taken to consider green areas; there was a call for 'densification with quality', meaning that green areas should be protected. In retrospect, cities have become more compact, but to a certain extent they have also encroached on green areas. The present Norwegian minister of the environment recently sent out a major political message, stating that urban green infrastructure has to be protected first, and that densification proposals may then be considered (Solheim 2010a, 2010b). We doubt that this is the priority in planning practice.

Neo-liberal Planning

The development of the compact city ideal coincided with a movement towards neo-liberal planning. This governance-oriented 'style' of planning differs from the traditional governmental system. This governmental system is based on a complex and often ambiguous set of norms and values related to political and administrative control, codes of professional conduct, governmental rules, democratic responsibility, public service ethics and the participation of affected groups (Christensen and Lægreid 2001). The transition to a neo-liberal mode of planning in Norway gave rise to project plans which were submitted by private actors. This change towards market-oriented planning also coincided with the New Public Management (NPM) reform, which took place in the public sector in the 1980s. NPM included a market orientation in public services. The neo-liberal movement and NPM brought to the fore an inherent rhetoric which aimed to provide administrative efficiency and economic freedom in a more impetuous way than a rigid command structure aimed at political control (Sager 2009). These two trends created tensions in planning with regard to the opportunities and constraints involved in balancing the contradictory elements of densification and the protection of green areas in the compact city model.

Planning in practice is heavily influenced by these trends. Today, development planning often begins with private project plans, which leads to a planning process of semi-formal meetings, negotiations and development agreements between developers and the planning authorities (Falleth et al. 2010). In effect, developers, private actors or public agencies are the real planners today, while the planning authorities guide the planning proposals, which are submitted towards political approval. The developers are responsible for the planning process, including plan-making, participation and public information. The authority is responsible for formal public hearings and political approval. A challenging question concerns the extent to which master planning frames project planning. The planning system allows development plans to be adopted which are not in accordance with higher-level plans. The links between these project plans and the overall master plan may therefore be weaker than their links to the building project (Falleth and Saglie 2011). The achievement of a good balance between development and the protection of green areas is difficult to achieve, as it depends on a large number of divergent elements and actors, including master planning, professional and political ideals, negotiations with developers, lobbying and participation and sanctions from national and regional authorities. Currently, there is a vital lack of democracy, transparency and political accountability in the world of planning. Deliberation in the planning process is based on collaboration and negotiation among certain actors, rather than being transparent and participatory. Falleth and Hansen (2009) performed a survey of developers who had submitted project plans for political approval and local organizations as representatives of the local community in the 145 largest municipalities in Norway. Table 12.1 illustrates the pattern of their interaction with politicians and the planning authority. While closed meetings were the most frequent kind before the planning process became regulated by the Planning and Building Act, public meetings now take place in the formal and regulated part of the planning process. Direct contact takes place throughout the entire planning process.

In total, 85 per cent of the municipalities in the survey had set up meetings with the developers in the planning process in order to administrate the project plans, while 90 per cent of the developers reported that they had participated in these kinds of meetings with the planning authority (see Table 12.1). The developers also reported having direct contact with the planning authority (43 per cent) and politicians (51 per cent). They rarely took part in open public meetings (2 per cent and 9 per cent). The local organizations had a more extensive pattern of interaction; they frequently had direct contact with the planning authority (73 per cent) and politicians (63 per cent). They also participated in public meetings (23 per cent and 26 per cent) and in closed meetings (20 per cent and 27 per cent) in which politicians and the planning authorities also participated. We believe that closed meetings between the developers and the planning authorities early in the planning process are especially influential in terms of framing the actual planning. Case studies have confirmed this argument (Nordahl 2006, Falleth et al. 2008, Netland 2008, Windju 2008). On the other hand, local organizations seem to rely on all available channels of communication in order to access the planning process, including meetings, direct contact and public meetings.

Green Arguments

Urban development has turned away from urban sprawl and towards densification in Norwegian cities and towns. This is in accordance with the compact city ideal. Planning practices have turned towards a form of project planning in which private developers are prominent actors. This change is evident in the change in the kinds of conflict which occur in urban planning. While just 3 per cent of development plans in Oslo, Bergen and Trondheim were in conflict with urban green areas in the city in 1987, this increased to 23 per cent in 2005 (Falleth et al. 2008). The minister

Table 12.1 Percentage of local organisations and developers reporting interaction with politicians and the planning authority during the development planning process (N=201/98)

	With local politicians		**With the planning authority**	
	Developers	Organisations	Developers	Organisations
Closed meetings	27	20	90	27
Open public meetings	9	23	2	26
Direct contact	51	63	43	73
N	98	201	98	201

for the environment called for a stronger will amongst local politicians to protect the green infrastructure. He seemed to be somewhat unsure about local politicians' willingness to protect green areas, and he added in a newspaper chronicle, that it 'also helps if people mobilize and get involved' (Solheim 2010a). Mobilization is a potentially powerful tool that comes into play for the first time late in the formal planning stage. This is often too late to change the result, however, due to investment in the planning process. At this stage, a change in direction is difficult, due both to the high transaction costs and path dependency resulting from the interaction in meetings and negotiations between developers and the planning authority, as well as with politicians.

While some interests can be socially excluded from important decision-making arenas, as often happens to green interests in the early stages of planning, they may also be subject to less visible forms of exclusion (Hanssen and Saglie 2010). A certain form of discourse can make some arguments seem valid while others are excluded, leading to cognitive closure. This cognitive form of closure can be observed in local mobilizations for the protection of green areas among planners and the political elites. Some of these mechanisms were revealed in 32 interviews conducted with local politicians, planners and local civil society organizations in Oslo, Bergen and Trondheim (Hanssen and Saglie 2010). The strong standing of the compact city argument was expressed by a local politician in the following way: 'there is high density insanity among planners. The more complaints the city receives from local actors the more successful they consider their high density strategy to be'.

There has been a local reaction against the narrow interpretation of the compact city in planning in practice. An umbrella organization for local organizations wishing to protect green areas in Oslo named the 'Network for green lungs' has been established. Local actors have learned that they have to adjust their arguments to fit the dominant planning discourse, which is the compact city discourse. Nonetheless, their argument that the protection of green areas is a legitimate representation of local interests has been widely perceived as a 'not in my backyard' (NIMBY) argument. These arguments have been interpreted as too narrow and too local. While a great deal of planning literature has underlined the importance of local knowledge, this does not seem to be legitimate in terms of the local protection of green areas, and particularly when the area is considered to be a resource-rich neighbourhood (Hanssen and Saglie 2010). This umbrella organization has therefore turned its arguments towards defending the green infrastructure in general and greenbelts in the city, not just green areas in small neighbourhoods. They argue that their opponents tried to define their arguments against development as protection of some local bushes or tiny woods. This is no longer valid when they argue that the green area is part of a green infrastructure. The difference in validity between these concepts is huge (Hanssen and Saglie 2010). By redefining their concerns from local to a larger scale, they have given green values public worth in the city and in the history of green planning in Oslo.

Planning a Green Oslo

Oslo is a green city with parks, trees, urban forests and waterways. Around two-thirds of Oslo's 454 km^2 is made up of green areas, and most of these green areas are covered by urban forests. Green values have been important throughout the history of planning in Oslo. 'Green fingers' and 'urban forests' have been the principles of planning since the first modern city plan in 1929. Green fingers were planned as corridors within urban built-up areas. These corridors are still important today as part of the current green infrastructure. The first protection of the urban forests took place in 1934, and a formal border against urban development was set by the national authorities in 1978. In 2009, the urban forests became protected by law (Byarkivet 2010).

Green values are also reflected in the Oslo municipal plan which was set out in 2008. The plan states that the City of Oslo 'manages natural resources in the city in a sustainable manner, cf Ecological Programme 2002–2014' (Oslo Municipality 2008: 14). Furthermore, the local government of Oslo said in their statement of approval of the plan that Oslo will be one of the world's most environmentally friendly cities. Fjord and the forest give Oslo a blue-green character to be preserved and strengthened. It was also stated in the master plan that the urban green structure should be protected or replaced with new green areas if any are lost. The chapter about the green-blue infrastructure states that it is important to ensure that no regulated recreation areas be used for buildings without a corresponding substitute, as well as that other important open areas have regulations in line with their usage.

Since the first elements of green planning in the master plan from 1929, green elements have been developed as a new planning discourse in Oslo. The first green plan for Oslo was adopted in 1993, and is currently under revision. The aim was to protect the green infrastructure, because the municipality had found that urban densification had threatened the green areas. This plan included parks, walking paths, waterways, green corridors and less-well maintained green areas in the city. A recent evaluation of the 1993 green plan concluded that the plan has provided an important argument in withstanding the pressure of development in green areas, but that it has not been a sufficient tool against the loss of green areas for development purposes (Oslo Municipality 2009).

There are no formal statistics which prove these statements. Parks, walkways and urban forests seem to be well protected in practice. This is less true for the rest of the green infrastructure, which is sometimes seen as 'just' fallow land or a playground. Statistics regarding the green infrastructure are not well developed. While there has been some research on the loss of green areas, it has been impossible to draw scientific conclusions owing to the wide range of sources and methods employed. The trend seems clear, however. One research project studied the loss of green areas between 1994 and 1996, and concluded that there has been a loss of 7.23 square metres per inhabitant in Oslo. This is equivalent in total to 640 football pitches (Thoren 2010). This means that the green areas which are being lost are not being replaced, as should occur according to the municipal plan. Generally speaking, there has been a decline in the number of recreational areas, particularly in the larger cities. A national report states that in Oslo, the total green area diminished from 70 to 67 square metres per inhabitant between 1999 and 2005 (The National Government 2005). According to the Oslo municipal plan, the green area per inhabitant is predicted to be 40 square metres in 2025 (Oslo Municipality 2008).

A wider and more multidimensional view of the usefulness of urban green areas has been adopted in national policy (Directorate for nature management 2003). The green structure is also viewed as being important for biodiversity, hydrology, aesthetics and pleasure, as well as recreation and health. The green infrastructure covers public and private areas; not only those designed for recreation, but all natural areas on private as well as public land (Ibid.: 8). In *PLAN*, particularly in the 1990s, some contributors were concerned with these kinds of barriers against the implementation of a densification policy. This fear has proven to be unnecessary. The compact city policy seems to have worked out. The trend since the beginning of the 1990s has been for higher population densities within major urban areas, although the Office of the Auditor General in Norway is dissatisfied with the government's work in implementing the compact city as set out in national policies (Office of the Auditor General 2007).

A green Oslo is regarded as being very important by the politicians, at least in policy documents and plans. Oslo has, however, chosen to have coarse and sketchy master plans. The argument has been that a legally comprehensive, detailed and binding plan for land usage is unsuitable as a tool for steering a rapidly growing and complex city. Pursuant to the new Planning and Building Act in 2008, the municipality had to present a legal master plan for land use in Oslo. In agreement with the Ministry of the Environment, the present master plan from 2008 is a temporary plan consisting of a list of previously adopted plans, including 16 area plans and 6,000 local development plans. This sketchy and non-binding master plan is still valid as a political expression of future development. We believe, based on the growing number of urban green organizations,[3] debates in the media[4] and planning research that project plans are often in conflict with the green structure (Thoren 2010, Falleth et al. 2008, 2010, Netland 2008, Windju 2008). This is not reflected in the professional debate in *PLAN*.

Arguments for and Against Sustainable Development

The two cases that have been chosen from Oslo are development projects which are in conflict with the urban green infrastructure. They have been chosen to illustrate the arguments and actions used in conflicts between urban development and the protection of green areas. The cases will illustrate the extent to which the different arguments for and against compact cities are put forward in planning in practice. The outcomes of the two cases are different. Geitmyra School Garden was protected as a school garden, while Tiedemann Playground is now a built-up area with small football pitch. However, local action caused the development scheme to be modified, and there is now a park in this area. The two cases reveal a significant number of arguments within controversial planning projects.

Geitmyra School Garden is a 1,000 m^2 public property located in the centre of Oslo. The school garden was established in 1909. It is located between the very busy Ring Road 2, the densely built-up 19th-century block structure and a graveyard. The garden is used by 14 local schools and 100 gardeners from all over the city. The

3 For example, Network for Protecting Green Areas and the Oslo River Forum, additional to traditional organizations such as recreational, environmental and nature protection organizations.

4 This included 93 articles about green lungs, and 18 about green infrastructure in Aftenposten 2006–10.

pupils cultivate vegetables and fruit and use the open space as a playground. The gardeners rent small allotments on an annual basis from the municipality.

There have been several development proposals put forward in the history of the school garden. In the 1990s, the city council decided to implement an active public land ownership strategy. This strategy included selling public land or developing public land for extensive use. Geitmyra was one of the areas which was regarded as having land which could potentially be used more intensively, in line with the compact city ideal. The Oslo municipality administration set up a scenario for Geitmyra which included 450 dwellings and an upgraded city park. The aim was to address the housing shortage in Oslo. The local protests made by environmental organizations, local neighbourhoods and schools were powerful. Their argument concerned the need to protect green areas for the children living nearby. As a result, the proposal was withdrawn.

In 2003, the national government decided to give high priority to creating more kindergartens, and municipalities were instructed to meet the demand for kindergartens. Oslo immediately started searching for areas to develop. One such area was Geitmyra Garden. The proposal was submitted in 2004. It included the development of a medium-sized kindergarten in part of the garden. The argument was that the proposal safeguarded both the need for a kindergarten and a school garden. The local protests were strong once again in relation to urban children's need for open spaces. The argument against development was extended from the need to protect green areas to the need to protect a cultural and natural heritage and biodiversity.

The local organizations in the area are strong, and have the capacity to mobilize and mix arguments from important discourses against development. They have also argued that the development plan conflicted with a high number of environmental objectives in the master plan and the green plan. These groups formed an ad hoc action group in 2004. The group 'Let the school garden survive – protect Geitmyra' received support from schools, environmental and cultural organizations, neighbourhood organizations and important individuals. They established a website,[5] arranged annual seminars and made several local arrangements. They used formal channels such as participation in planning (open meetings, participation, writing hearings) as well as more informal channels (networking, direct meetings, lobbying and media).

The local mobilization committee proposed a cultural institution in the garden as an alternative to the kindergarten. It included a food centre for children to promote public health, food production and agricultural education. In 2009, the minister for agriculture took part in the 100th anniversary of the garden. The minister was willing to support the food centre. Later that year, the minister for the

5 http://www.vern-geitmyra.no (accessed: 3 July 2010).

environment also supported this idea. The development plan was withdrawn and the food centre was established in 2010.[6]

Tiedemann Playground was a 12,000 m^2 private property within the larger urban transformation area of Ensjø. Ensjø is a heterogeneous, low-density area, with a diverse range of land uses, including business, dwellings, parking spaces and green semi-public spaces. The development programme for Ensjø was adopted by the city council in 2004, and the aim was that Ensjø should be transformed into a residential area with 5,000–7,000 dwellings. The main arguments for the development project are densification and closeness to the public transport network, and that green areas and housing would be combined in order to secure sustainable development. These arguments concur with the compact city ideal. The Tiedemann area was originally designated for recreation, sports and parking in the master plan. However, the area *de facto* was neither a park nor a recreation space. The developers argued that the area was fallow land, and an appropriate area for densification. The developer submitted a development plan in 2005, and the plan was adopted by the city government later that year.

This proposed plan also met with strong local resistance, which led to adjustments being made to the plan, including open areas and two small football pitches. Today, the area is built up, and contains 190 flats. The local government promised to make a park in the Tiedemann area. This plan was submitted for public hearing in the summer of 2010. Local organizations also protested against the plan for a park, because they wanted a playground for children and not a park. Their argument was that the proposal was against the national guidelines to protect children and their rights to sustainable playgrounds and green areas.

Table 12.2 provides an overview of the arguments and outcomes in the two areas. The two cases have both similarities and differences in terms of their institutional settings. Both were green areas used by children. Both green areas were set aside as green areas in the municipal plan and in the green plan. The cases differ with respect to other institutional settings. Geitmyra School Garden is a public property, administrated by an old institution and which has been used by the local schools for many decades. The Tiedemann Playground was privately owned, and while it was in use as a playground, the area was not administrated by any institution in the area. The green areas in both places were regarded as fallow land by some actors.

The arguments for and against development vary. The Tiedemann case addressed the compact city ideals which have been interpreted as densification and the upgrading of fallow land. The case of Geitmyra focused on the need for kindergartens, densification and protection of playgrounds. The local arguments against development, in both cases, addressed the protection of the green infrastructure, recreational areas and playgrounds for children in urban areas. None of these arguments addressed efficient land use or the transport development

6 There is also proposal for road enlargement that will affect the garden. The action group has been invited to take part in the planning process.

Table 12.2 Arguments and outcomes in Tiedemann Playground and Geitmyra School Garden

	Tiedemann Playground	Geitmyra School Garden
Arguments for development	Densification, public transport accessibility, housing shortage, fallow land, sustainable development	Lack of kindergartens, vacant land, combining kindergarten and school garden
Arguments against development	Children, playground, recreation, lack of open space in inner city	Children, playground, recreation, environmental and cultural values, education, public health, lack of open space in inner city
Outcome	Developed for residential use in 2008 Plan for a park and playground in 2010	Housing development scheme was withdrawn in the 1990s Programme withdrawn 2009 Centre for food and culture established in 2010

pattern. Despite many similar arguments, the outcomes of the two cases were different. Tiedemann was developed, while Geitmyra was not.

So why was development in Geitmyra stopped? The most striking feature of this case is that Geitmyra has a strong local identity. A high number of local institutions mobilized against its development. This stopped the housing development scheme at first, and later the kindergarten plans. The local activists addressed a high number of political priorities, such as the need for a playground for children living in urban areas, the environment, recreation, local culture and public health. They addressed the institutionalized history of the school garden, and created publicity using several channels. The well-organized local community utilized their social networks and invited political celebrities to collaborate with them, which gave them good access to the media. They also managed to combine current political trends with their local identity by suggesting a food and culture centre – and thereby a place for innovation. They managed to put forward ideas by combining political and professional arguments for protection in the form of a plan for a cultural institution, and they generated political pressure by engaging the ministers for agriculture and the environment in their efforts.

The Tiedemann area had a small but active number of local organizations, but no local protest committee. The protests which did occur were based on fewer political arguments. In addition, importantly, the actors' social networks were less extensive. They never managed to threaten the development. However, they managed to modify the development plan, by including a small football pitch within the built-up area. In addition, the status of the two areas was different: while Geitmyra was a well-established school garden with a long history, the Tiedemann Playground was argued to be just 'fallow land', and it was therefore easier to argue for construction. We argue that the outcome of the two cases may be explained as a result of the differences between the local institutions and networks, mixed with a high number of similar arguments against development. Combining arguments, local institutions, networks and the local identity seems to be of crucial importance when comparing these two cases.

These cases also illustrate a gap in the emphasis of sustainable development between the professional debate in *PLAN*, which focuses on efficient land use and the transport development pattern, and the local debate, which addresses the loss of green areas. The argument for efficient transportation and dense land usage has been of less importance for local actors. On the other hand, densification is of high importance for developers and planners. The arguments against development involve children, playgrounds, biodiversity and green areas. These two cases also illustrate that the compact city ideal, in practice, seems to be related to densification rather than the protection of the green infrastructure. The protection of green areas seems to be left to local actors.

Conclusion

This chapter describes the compact city ideal, project planning and the protection of urban green areas. We have investigated the extent to which the protection of the urban green infrastructure is regarded as important in urban development. While the compact city ideal has underlined the need to achieve a balance between the protection of urban green structures and the implementation of densification, the protection of green areas seems to be less evident in planning practice and the professional debate. The professional discourse in *PLAN* has been dominated by transport and land use. Less focus has been giving to the protection of the green infrastructure. In this respect, green interests seem to be paid very little attention among professionals. This lack of emphasis may have affected both private and public planners' attitudes, as they seem to be less interested in this issue and consequently fail to recognize its importance in the public debate over green areas, leading to a gap between planning practice and planning research.

By contrast, local actors pay a great deal of attention to the protection of green areas. This has been illustrated using two contested development schemes, which both threatened the green infrastructure in Oslo. We have described the bottom-up character of present-day planning practice, which may well challenge the overarching goals laid out in both national guidelines and higher order plans, such as master plans. This is also evident in Oslo, where higher order plans and outspoken policy goals to protect the green structure are not sufficient to protect green areas. There is a constant flow of development proposals which threaten green areas. The present planning practice thus favours the 'development' side. However, local resistance against development proposals has protected at least some green areas. Local mobilization seems to be crucial for the protection of urban green areas.

National policy favours the compact city as ideal for sustainable urban development. Developers address the densification part of this policy, while local actors address the protection of green areas. These actors, therefore, interpret the concept of sustainable development differently and strategically. National policy calls for both densification and the protection of green areas. The use of arguments is not surprising. We are, however, surprised to find that the professional debate also ignores or pays very little attention to the protection of urban green areas as an important element in the compact city ideal. This has resulted in an apparently shared understanding regarding what urban sustainable development is between developers and professionals.

We also studied the strength of arguments for the protection of the green urban infrastructure. The cases used illustrate that formal planning, in the form of master plans and green plans, is not sufficient to protect the green infrastructure against urban development. Both cases were planned as open green areas, but development schemes were proposed and, in one area, implemented. Local mobilization managed to stop development in the second case. As it was 'just a playground', Tiedemann was interpreted as being fallow land. Being 'just a playground' needs to be combined with stronger and more in-depth arguments including policy, history and identity, as in Geitmyra. The different kinds of interaction between environmental, cultural and social policy arguments, local identity and strong local organizations can be used to explain the different output of the two cases. Local actors' capacity to translate local issues into the language of national policy discourse seems to be the most important factor.

The ideal of the compact city is strong within the planning discourse on urban sustainability. High density is one important element in the compact city ideal, which is often a part of urban development projects. Urban growth requires housing development. Politicians, planners and developers must therefore develop a joint focus with regard to the compact city in order to promote high density in development projects, despite not sharing values and knowledge. Development plans have an increasingly autonomous role due to project plans in the neo-liberal planning system. This has strengthened the tendency to address densification issues while neglecting green areas in urban development. The continuous flow of almost autonomous private development proposals does indeed constitute a threat to the urban green structure. 'Green arguments' are weak, particularly in areas where green areas seem to be in extensive use or defined by many as fallow land, thereby meeting the requirements for development. We find it unsatisfactory that the protection of urban green areas seems to depend on local actors and their capacity to act.

Acknowledgements

This chapter is based on two research projects entitled 'From participation as a plus factor in government to participation as a strategy in governance' and 'Conditions for sustainable management of urban recreational landscapes. May governance and accessibility be valuable tools?'. They are funded by the Norwegian Research

Council research programmes entitled *Democracy, Regionalism and Governance* and *Environment 2015*.

Bibliography

Breheny, M. 1992. *European Research in Regional Science*. London: Pion, 138–59.

Byarkivet 2010. *Markahistorie*. Available at: http://www.byarkivet.oslo.kommune.no/OBA/tobias/tobiasartikler/t3981.htm [accessed: 16 July 2010].

Børrud, E. 2005. *Bitvis byutvikling: Møte mellom privat eiendomsutvikling og offentlig byplanlegging*. Doctoral thesis. Oslo: School of Architecture.

Christensen, T. and Lægreid, T. 2001. New public management – undermining political control?, in *New Public Management: The Transformation of Ideas and Practice*, edited by T. Christensen and T. Lægreid. Aldershot: Ashgate, 93–119.

Directorate for Nature Management. 2003. *Grønn by: Arealplanlegging og grønnstruktur*. Trondheim: Directorate for Nature Management.

Falleth, E.I. Hanssen, G.S. and Saglie, I.L. 2008. *Medvirkning i byplanlegging i Norge*. NIBR – report 2008:37. Oslo: Norwegian Institute for Urban and Regional Research.

Falleth, E.I. and Hanssen, G.S. 2009. Deltakelse og innflytelse i byplanlegging – en undersøkelse av reguleringsplanprosesser. *Kart og Plan*, 69(1), 46–56.

Falleth, E.I., Hanssen, G.S. and Saglie, I.L. 2010. Challenges to democracy in a market oriented urban planning in Norway. *European Planning Studies*, 18(5), 737–54.

Falleth, E.I. and Saglie, I.L. 2011. Democracy or efficiency: Contradictory national guidelines in urban planning in Norway. *Urban Research & Practice*, 4(1), 58–71.

Groenne lunger. 2010. *Utbyggingssaker i konflikt med groenne lunger.* Available at: http://www.gronnelunger.no/documents/saker.php [accessed: 17 July 2010].

Hanssen, G.S. and Saglie I.L. 2010. Cognitive closure in urban planning. *Planning Theory and Practice*, 11(4), 499–521.

Hanssen, G.S. 2010. Ensuring local community interests in market-oriented urban planning? The role of local politicians. *Environment and Planning C – Government and Policy*, 28(4), 714–32.

Jenks, M., Burton, E. and Williams, K. (eds) 1996. *The Compact City: A Sustainable Urban Form?* London: E&FN Spon.

Kalbro, T., Lindgren, E. and Røsnes, A. 2010. Nær utakt plan – og bygningslovsreformer i Norge og Sverige. *Kart og Plan*, 70(1), 27–45.

Kleven, T. 2011. *Norsk kommuneplanleggings historie fra 1965 til i dag*. Trondheim: Tapir Akademiske Forlag.

Le Corbusier, F.E. 1925/1971. *Urbanisme*, translated as *The City of Tomorrow and its Planning*. Cambridge, MA: MIT.

Ministry of the Environment. 1992. *Den regionale planleggingen og arealpolitikken.* St. meld. No. 31 (1992–93). White Paper to the Parliament. Oslo: Office of the Prime Minister and the Ministries.

Ministry of the Environment. 1996. *Miljøvernpolitikk for en bærekraftig utvikling.* Dugnad for framtida. St. meld. No. 58 (1996–97) White Paper to the Parliament. Oslo: Office of the Prime Minister and the Ministries.

Ministry of the Environment 2001. *Bedre miljø i byer og tettsteder.* St. meld. No. 23 (2001–02) White Paper to the Parliament. Oslo: Office of the Prime Minister and the Ministries.

Netland, M.E. 2008. *Rom for dissens? Medvirkning og governance i privat planlegging.* Master's thesis at Department of Landscape Architecture and Spatial Planning. Aas: Norwegian University of Life Sciences.

Nordahl, B. 2006. *Deciding on Development: Collaboration between Markets and Local Governments.* Doctoral thesis 2006(96). Trondheim: Norwegian University of Science and Technology.

Norway Statistics 2010. *Population Statistics.* Available at: www.ssb.no/statistikkbanken [accessed: 22 December 2010].

Norway Statistics 2011. *KOSTRA.* Available at: www.ssb.no/kostra [accessed: 19 February 2011].

Næss, P., Næss, T. and Strand, A. 2009. *The Challenge of Sustainable Mobility in Urban Planning and Development in Oslo Metropolitan Area.* TØI report 1024/2009. Oslo: Norwegian Centre for Transport Research.

Næss, P. 1992. *Natur – og miljøvennlig tettstedsutvikling (NAMIT).* Final Report. Oslo: Norwegian Institute for Urban and Regional Research.

Office of the Auditor General 2007. *Bærekraftig arealplanlegging og arealdisponering i Norge.* Document 3:11(2006–07). Oslo: Fagbokforlaget.

Oslo Municipality 2008. *Oslo Municipality Plan.* Available at: http://www.byradsavdeling-for-finans-og-naring.oslo.kommune.no/getfile.php/byr%C3%A5dsavdeling%20for%20finans%20%28FIN%29/Internett%20%28FIN%29/Dokumenter/Kommuneplan%20vedtatt_2008_uten%20kart.pdf [accessed: 19 February 2011].

Oslo Municipality 2009. *Evaluation of the Oslo Green Plan.* Available at: http://www.plan-og-bygningsetaten.oslo.kommune.no/getfile.php/plan-%20og%20bygningsetaten%20%28PBE%29/Internett%20%28PBE%29/Dokumenter/Filer%20utlagte%20saker/2009/Evaluering%20av%20gjeldende%20grontplan.pdf [accessed: 20 December 2010].

Røsnes, A. 2005. Regulatory power, network tools and market behaviour: Transforming practices in Norwegian urban planning. *Planning Theory and Practice*, 6(1), 35–51.

Sager, T. 2009. Planners' role: Torn between dialogical ideals and neo-liberal realities. *European Planning Studies*, 17(1), 65–84.

Solheim, E. 2010a. *Stå på for grønne lunger.* Chronicle in Aftenposten 2.11.10:6.

Solheim, E. 2010b. *Opening Speech, Planmøte 2010.* Available at: www.boby.no/aktuelt/norsk_planmote_2010 [accessed: 15 November 2010].

The National Government 2005. *Fremtidens byer.* Available at: http://www.regjeringen.no/nb/sub/fremtidensbyer/forside.html?id=551422 [accessed: 7 December 2010].

Thoren, K.H. 2010. Grønnstruktur i by. Hvordan takle endringene? *PLAN* (3/4), 30–35.

World Commission on Environment and Development 1987. *Our Common Future.* Oxford University Press.

Windju, M. (2008) *Medvirkning i privatinitiert planlegging.* Master's thesis, Department of Landscape Architecture and Spatial planning. Aas, Norwegian University of Life Sciences.

Index

For Product Safety Concerns and Information please contact our EU representative GPSR@taylorandfrancis.com Taylor & Francis Verlag GmbH, Kaufingerstraße 24, 80331 München, Germany